T0192635

Introducing
Genetics

SECOND EDITION

Introducing **Genetics**

SECOND EDITION

ALISON THOMAS

Routledge
Taylor & Francis Group

LONDON AND NEW YORK

Vice President: Denise Schanck
Senior Editor: Elizabeth Owen
Assistant Editor: David Borrowdale
Production Editors: Ioana Moldovan and Georgina Lucas
Layout: Phoenix Photosetting
Cover Designer: Susan Schmidler
Copyeditor: Ray Loughlin
Proofreader: Susan Wood

First published 2015 by Garland Science

2 Park Square, Milton Park, Abingdon, Oxfordshire OX14 4RN
52 Vanderbilt Avenue, New York, NY 10017

Routledge is an imprint of the Taylor & Francis Group, an informa business

First issued in paperback 2019

ISBN 978-0-8153-4509-1 (pbk)

Library of Congress Cataloging-in-Publication Data
Thomas, Alison.
 Introducing genetics : from Mendel to molecules / Alison Thomas.
-- Second edition.
    pages cm
 ISBN 978-0-8153-4509-1 (paperback)
1. Genetics. I. Title.
 QH430.T48 2015
 576.5--dc23
                            2014041220

Preface

As with the first edition, my main aim for the second edition of *Introducing Genetics* was to write a concise and easy-to-read introduction to the three key areas of genetics: Mendelian, molecular, and population. I have written for first year students and those studying to become biology majors, to give you a thorough grounding in genetics and prepare you for further study. The big general genetics textbooks can be imposing, especially if you have no prior genetics knowledge, but *Introducing Genetics* is written in a style that will engage, encourage, and inform. If you have previously struggled with genetics this book could change things for you. It first establishes the principles of Mendelian inheritance and the nature of chromosomes, before tackling quantitative and population genetics. The final three chapters introduce the molecular mechanisms that underlie genetics, including the techniques responsible for the current genetic revolution. As genetics affects all living organisms, I have taken examples from insects, plants, animals, and humans. The end of each chapter has questions that are carefully designed to help build your confidence in your developing understanding of genetics. I hope after reading my book you will have both the core knowledge and, most importantly, the confidence to tackle genetics in any context. Good luck with your studies!

Alison Thomas
November 2014

Acknowledgments

I have greatly appreciated the support and encouragement that I have received from the editorial and production team at Garland Science during work on this second edition. I would particularly like to thank Liz Owen, Senior Editor, and Georgina Lucas, Senior Production Editor. Special thanks are also due to Ray Loughlin, Copy Editor, Ioana Moldovan, Production Editor, and David Borrowdale, Assistant Editor. I would like to thank Dawn Hawkins and Helen Roy for their valuable discussions and comments on the manuscript during preparation of the first edition.

The author and publisher would like to thank the following external advisers and reviewers for their suggestions and advice in preparing *Introducing Genetics, Second Edition*.

John Acord, London South Bank University, UK; Samantha Alsbury, University of Greenwich, UK; Shazia Chaudhry, University of Manchester, UK; Robert Fowler, San José State University, USA; Emma Ghaffari, Brunel University, UK; Lynne Hardy, City of Sunderland College, UK; Sarabjit Mastana, Loughborough University, UK; Claire Morgan, Swansea University, UK; Cynthia Wagner, University of Maryland, Baltimore County, USA.

Contents

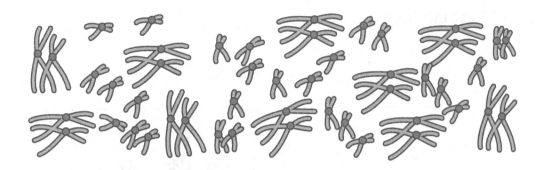

Introduction

Each of us starts life as a single fertilized egg that develops, by division and differentiation, into a mature adult consisting of roughly a hundred trillion (10^{14}) cells. Each cell is specialized for a particular function, such as a muscle cell able to contract or a neuron programmed to conduct a nerve impulse. The information that guides this carefully orchestrated development and maintains a fully functioning adult is contained in an estimated 21,000 genes, and is itself stored within each and every cell. The discipline of genetics is concerned with attempting to understand the nature of such information, how it is transmitted from generation to generation, and how it is stored, expressed, and regulated.

Genetics, as a scientific discipline, is a twentieth-century development, although the recognition of the principle of heredity is clearly much older. The development of agriculture and farming, from around 8000 BC, was only possible because people realized that desirable (and undesirable traits) could be passed to successive generations. The domestication of animals and cultivation of plants was achieved by selecting those genetic variants having the characteristics desired by farmers. For example, modern wheat, *Triticum aestivum*, has been produced through selective breeding from the smaller-eared *Aegilops speltoides* (Figure 1.1).

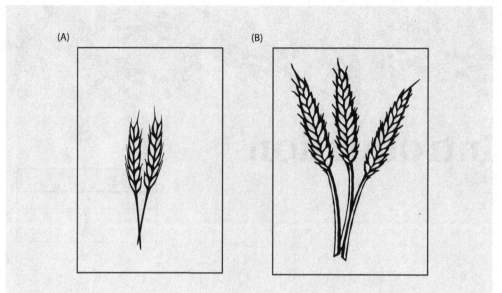

Figure 1.1 The development of modern wheat. (A) *Aegilops speltoides*. (B) *Triticum aestivum*.

Over the centuries there have been many attempts to explain inheritance and the link between succeeding generations, but it was Gregor Mendel (1822–1884) who proposed the scientific theory we accept today. Between 1856 and 1863, he performed a series of carefully planned and executed breeding experiments with pea plants in the monastery garden of the Augustinian Abbey of St Thomas at Brno, now in the Czech Republic, where he lived as a monk. He concluded that the patterns of inheritance he observed between succeeding generations were only consistent with a **particulate** explanation for inheritance. His theory involved **hereditary factors**, each controlling a separate trait, which passed unchanged from parent to offspring during reproduction.

Mendel published an account of his work in 1866. Unfortunately, the intellectual climate of that time was not ready to accept this challenge to the then widely held belief in a blending basis to heredity and his work went largely unnoticed. It was not until 1900, 16 years after Mendel's death, that the significance of his findings was fully appreciated. Three botanists, Hugo de Vries (1848–1935), Carl Correns (1864–1933), and Erich von Tschermak (1871–1962), each independently rediscovered Mendel's paper when analyzing the results of their own similar breeding experiments, and immediately recognized the importance of his work. In the intervening years since Mendel's experiments there had been progress in other complementary areas of biology. Improvements in microscopy had revealed the presence of discrete chromosomes in the cell nucleus, and it was quickly realized that the transmission of chromosomes during cell division and reproduction exactly paralleled the behavior of Mendel's hereditary factors.

A particulate basis to heredity was now plausible and the science of genetics was born.

Genetics is a huge and varied discipline, which approaches inheritance from various aspects – molecular and cellular; in individuals and in populations. Before considering different aspects of the hereditary process in detail, a simplified overview of key principles will be given, with the aim of introducing some initial vocabulary and a conceptual framework in which to fit ideas as they are considered in subsequent chapters.

1.1 Key concepts in genetics

The physical site of potential heredity in **eukaryotic** organisms (animals, plants, and fungi) is the central **nucleus** of each cell. In **prokaryotic** (bacterial) cells the genetic material is in a less-well-defined central area – the **nucleoid**. In both eukaryotes and prokaryotes, the genetic material is **DNA**, an abbreviation for deoxyribonucleic acid. **RNA** (ribonucleic acid) is another nucleic acid involved in the transfer of information from the nucleus to the cytoplasm and in its expression. Occasionally, in a few viruses, RNA functions as the genetic material. Nucleic acids, along with proteins, carbohydrates, and lipids, make up the four major classes of biomolecules found in living organisms.

Within the eukaryotic cell nucleus the genetic material is distributed between a number of **chromosomes**. Each chromosome consists of a single linear DNA molecule complexed with protein. By contrast, each bacterial cell contains a single, circular DNA molecule. Different regions of a chromosome represent different hereditary units (**genes**). A gene, therefore, can be defined in functional terms as an informational storage unit that can be replicated, expressed, and regulated. From a chemical perspective it is simply a piece of DNA.

Chromosomes are only easily visualized when nuclei are dividing as a prelude to cell division – during the processes of **mitosis** and **meiosis**. Mitosis occurs when the genetic material is exactly duplicated and a complete copy distributed to each of two new nuclei during normal cell growth. Meiosis refers to the process that halves the amount of genetic material in a cell, generally when gametes are forming prior to sexual reproduction. At such times of nuclear division, the DNA and associated protein of a chromosome tightly coil and condense, giving rise to the characteristic image of a chromosome – two **chromatids** connected at the **centromere** (Figure 1.2). Between cell divisions, chromosomes exist in an uncoiled state, known as **chromatin**, and individual chromosomes are indistinguishable (Figure 1.3).

The chemical building blocks of DNA are called **nucleotides**, of which there are four types differing in their nitrogenous base component: adenine, guanine,

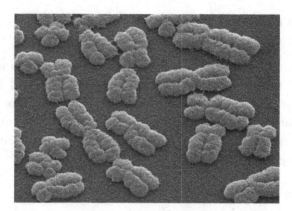

Figure 1.2 Scanning electron micrograph of human mitotic chromosomes. Note how each chromosome consists of a pair of chromatids held together at a constriction point – the centromere. Courtesy of Science Photo Library.

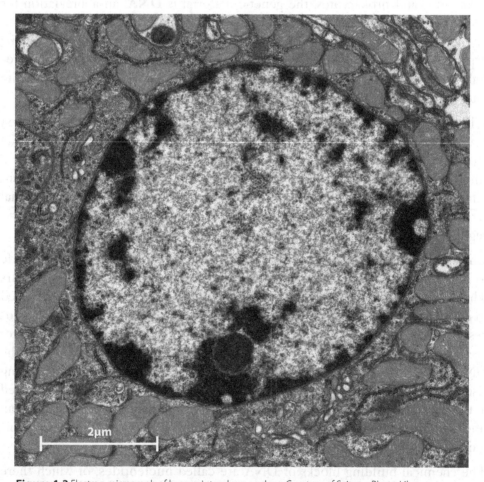

Figure 1.3 Electron micrograph of human interphase nucleus. Courtesy of Science Photo Library.

cytosine, and thymine nucleotides. The key to a gene's storage of hereditary information lies in the precise sequence of these four nucleotides in a gene. Expression of a gene's stored information leads directly to the synthesis of a protein. The coded information within DNA is first transferred, during a process called **transcription**, into an RNA molecule; specifically, a messenger RNA molecule. This RNA molecule then associates with a cellular organelle, the **ribosome**, where it directs synthesis of the encoded protein molecule, during a process called **translation** (Figure 1.4). Proteins perform a range of diverse roles within living organisms. Many are enzymes, catalyzing biological reactions. Others perform roles in the structural, immunological, nervous, or hormonal systems. As genes encode proteins, one goal of the study of inheritance is to understand how the great diversity of inherited traits can be related to the working of proteins.

The study of genetics, therefore, involves all living organisms and encompasses all levels of biological organization, from molecules to populations. In seeking to understand ways in which genes determine traits, a range of approaches is adopted, producing three main branches of genetics: **transmission, molecular,** and **population**. The earliest, and "classical," approach is the study of **transmission genetics**, also often referred to as **Mendelian genetics**. Insights into genetic principles are gained from studying patterns of inheritance from parents to offspring over several generations. **Molecular genetics** has had the greatest

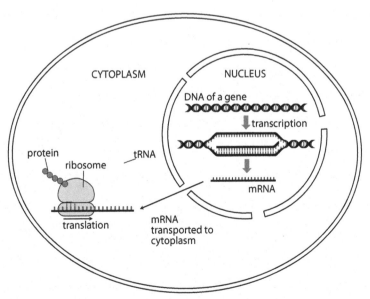

Figure 1.4 DNA to protein: a summary of the key stages involved in expression of a gene's stored information. DNA is transcribed into mRNA in the nucleus. The mRNA is transported to the cytoplasm where it is translated into protein.

impact on genetic knowledge over recent decades and has led to the development of the discipline of **recombinant DNA technology**, where genes controlling specific traits are identified, sequenced, cloned, and physically manipulated – all with profound implications for medicine, agriculture, and bioethics. **Population genetics** analyses the variation among individuals present within a population and considers mechanisms that promote the maintenance of certain kinds of variation at the expense of others. Such information provides possible insights into evolutionary processes. It also enables us to make predictions about future variation, which has implications, for example, for biological conservation programs.

Together, these various approaches to the study of genetics have produced one of the most advanced scientific disciplines. The aim of this book is to give a straightforward and informative introduction to this diverse and ever-expanding subject, in our current "age of the genome."

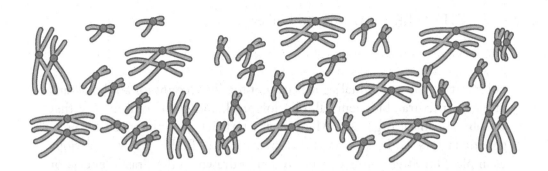

Monohybrid Inheritance

We are all the same – yet different! Look around any crowded high street on a Saturday morning. Everyone is unmistakably human – there are similarities. Yet at another level there is tremendous variability – in height, in skin color, in face shapes, in build. Some people might have brown hair, while others have blonde; some curly, others straight. If you had a sophisticated molecular biology testing kit to hand you would find variations in the activity of particular enzymes, in blood groups, and in many other "hidden" characters. To account for all this variability within a species we can examine the actions of genes. Many of the differences and similarities between individuals can be understood by looking at the transmission patterns of genes from parents to offspring. Certain "rules of inheritance" exist. Understanding these rules can explain the *status quo* in any generation and enable us to make predictions about the expression of traits in future generations (important in many situations, such as family planning, agriculture, livestock farming, and conservation).

Recognizing the basic rules of inheritance marked the beginning of the modern discipline of genetics and is the work for which Gregor Mendel is famed. It is important, though, at the start of any study of genetics to point out that an individual's genes cannot totally account for all the details of each structural, biochemical, physiological, and behavioral feature that we observe in an organism. Many characteristics are influenced by non-genetic factors (i.e. by aspects of the environment). Body mass is an obvious example. Our attention is constantly being drawn to the drastic effects of overeating and little exercise on human body mass! More controversial are environmental and other influences on behavioral features and personality. It is essential that we recognize that there are other, non-genetic influences acting on organisms and, consequently, that we do not view genetics in a totally deterministic way. Individuality cannot be wholly accounted for by an organism's genes. How we may assess the relative contribution of genes and environment to a given trait is discussed in Chapter 9.

The aim of this chapter is to introduce:

- Some fundamental principles of genetic inheritance

- The terminology associated with these key principles

2.1   Key principles of genetic inheritance

These principles explain the expression of a given trait in an individual in one generation and enable predictions to be made about future generations. The best context in which to understand them is that of breeding experiments. One species in which there has been intensive breeding is the tomato, *Solanum lycopersicum*. We will consider one characteristic – the color of its fruit. There are two possible colors – red and yellow. Most commercially grown varieties yield red fruit, but there are also yellow varieties. Two main questions will be addressed in this section:

1. What color fruits are produced if red- and yellow-fruiting varieties are crossed?

2. How can we explain the consistency of fruit color from generation to generation in the different varieties?

First, we will consider the consequences of crossing a red- and a yellow-fruiting variety. In a breeding experiment of this nature, when the genetic basis of one character is being investigated (a **monohybrid cross**), there is a set procedure:

1. **True- or pure-breeding** parents are used. This means that each parent comes from a variety in which (a) all members express the same characteristic, and (b) breeding within a variety produces offspring that *all* show the same characteristic as each other and as their parents.

2. **Cross-fertilization** is performed between the pure-breeding parents. This produces the F_1 (**first filial**) **generation**. The traits expressed by the F_1 individuals are observed.

3. The F_1 seeds are planted and the resulting individuals crossed among each other to produce the F_2 (**second filial**) **generation**. The traits expressed by the F_2 individuals are observed.

4. Large numbers of crosses are performed to produce large numbers of offspring, so that any statistically significant trends can be recognized.

5. **Reciprocal** crosses are performed (i.e. there are two sets of parental crosses). In the first set, crosses are between males showing one trait (here, red fruiting) and females the other trait (here, yellow fruiting). In the other set of crosses, the traits are reversed (here, yellow-fruiting males crossed with red-fruiting females) This strategy eliminates gender-related differences.

6. Precautions are taken to ensure that only the desired fertilizations take place. For example, with plants, the flowers are wrapped in muslin bags to prevent 'foreign' pollen gaining access to the stigma. Box 2.1 presents an overview of the basic reproductive process in higher plants.

BOX 2.1 REPRODUCTION IN HIGHER PLANTS

The flowers of most higher plants are hermaphrodite; that is, they contain both male (the stamen) and female (the carpel) reproductive organs. Sexual reproduction involves pollen, containing the male gamete, being transferred by wind or insects to a receptive female structure (the stigma). A tube then grows out from the pollen grain and through the female tissues. This provides a pathway for the male gamete to reach the eggs in the ovary. Generally, many pollen grains land on one stigma. Thus, many pollen tubes grow towards many eggs. The eventual result is many fertilizations and so many seeds within one fruit – as in the tomato.

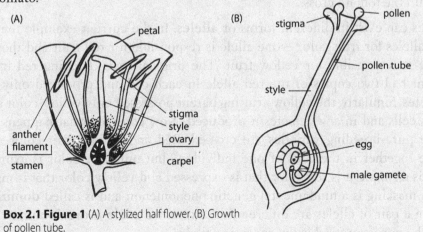

Box 2.1 Figure 1 (A) A stylized half flower. (B) Growth of pollen tube.

Figure 2.1 Investigating the inheritance of tomato fruit color (note that the F_1 red fruiters are interbred to produce the F_2 generation).

Figure 2.1 shows the results of performing a standard monohybrid cross as outlined above to investigate the genetic basis of fruit color in tomatoes.

2.2   A genetic explanation of the monohybrid cross

The most striking feature of this monohybrid tomato cross is probably the F_1 result. All F_1 plants produce red fruits. This is true whichever way the cross is carried out – whether the parental pollen comes from the red- or yellow-fruiting variety. Yet when the F_1 plants are self-fertilized, yellow-fruiting plants are again present in the next generation. These results illustrate some key principles of genetic inheritance.

Within the chromosome set of the tomato there is a gene carrying the instructions for fruit color. A normal working cell of the tomato contains two copies of the gene for fruit color – one copy will have come via the female gamete and the other via the male. As a prelude to sexual reproduction an organism produces gametes. The cellular process of **meiosis** (see Chapter 5) ensures that each gamete contains just one copy of each gene so that at fertilization there is once again two copies per trait and normality within cells is maintained. We can now explain the tomato cross.

Genes can exist in different forms or **alleles**. In the current example there are two alleles for fruit color – one allele is responsible for red fruit and the other allele is responsible for yellow fruit. The original pure-breeding red-fruiting parent had two copies of the red allele in each cell and produced only "red" gametes. Similarly, the yellow-fruiting parent possessed only yellow color alleles in its cells and in any gametes it produced. Now consider what happens when these pure-breeding varieties are crossed. Red and yellow fruit color alleles come together in the cells of one individual, but only one color is expressed. In this example it is red color that is expressed and yellow color that is masked. Such masking is a fundamental genetic phenomenon and is called **dominance**. When a pair of alleles are different, the expressed allele is the **dominant** allele and the non-expressed is the **recessive** allele.

You may have noticed that dominant expression occurs when alleles are the same (both red) or different (one red and one yellow), but recessive expression only occurs in the presence of two identical recessive alleles. This observation introduces two more genetic terms. A **homozygote** refers to an individual that possesses two identical alleles. When the alleles are different, an individual is said to be a **heterozygote**. Hence, we have the terms **homozygous dominant**, **heterozygous dominant**, and **homozygous recessive**. These three terms represent different **genotypes**.

Genotype and **phenotype** are two other commonly used genetic terms. Genotype refers to the particular gene(s) associated with a given trait and phenotype refers to the physical characteristic that results from expression of these gene(s). Thus, individuals with a homozygous recessive genotype produce the yellow-fruiting phenotype. Genotype, therefore, determines phenotype. Sometimes these terms are used in a broader sense, where phenotype refers to the sum of all characters that an individual possesses (i.e. structurally, biochemically, physiologically, and behaviorally) and genotype alludes to the full complement of an individual's genes.

The preceding paragraphs explain the results of the tomato cross between red and yellow fruiters, and introduce some genetic terminology, summarized in Figure 2.2. Section 2.3 presents a shorthand way of representing crosses.

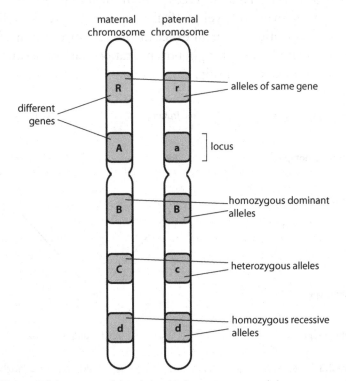

Figure 2.2 A summary of the relationship between genes and chromosomes.

2.3 Representing genetic crosses

Genes are represented by letters of the alphabet – upper case for the dominant allele and lower case for the recessive allele. The letter chosen is generally the first letter of the dominant phenotype. Thus, **R** can be used to represent the allele for red fruit and **r** the allele for yellow fruit. The production of the F_1 and F_2 generations, using symbols for alleles, is represented in Figures 2.3 and 2.4.

When representing the results of genetic crosses we tend to express relative numbers of different phenotypic classes as **ratios**. Offspring ratios often give valuable clues to an individual's genotype and to which of two alternative characters is dominant.

A few comments are needed about representing fertilizations where an individual produces more than one type of gamete, as in the F_1 red-fruiting plants. The lines linking gametes from plant 1 with those from plant 2 show all the possible fertilizations and assume an equal chance or probability of all events. Consider plant 1: an **R**-bearing gamete can combine with an **R** or **r**-bearing gamete from plant 2. Over a large number of fertilizations we could expect 50% of each type of fertilization – 50% homozygous red plants (**RR**) and 50% heterozygotes (**Rr**). Likewise, the **r**-bearing gamete has an equal chance of combining with an **R**- or **r**-bearing gamete from plant 2, so producing 50% heterozygotes (**Rr**) and 50% yellow homozygotes (**rr**). This produces a theoretical phenotypic ratio of three red-fruiting plants to every one yellow-fruiting plant. Rarely, though, do the different types of fertilization occur in exact theoretical proportions. So when we analyze results, we are looking for phenotypic ratios that approximate to 3 : 1 or some other informative relationship.

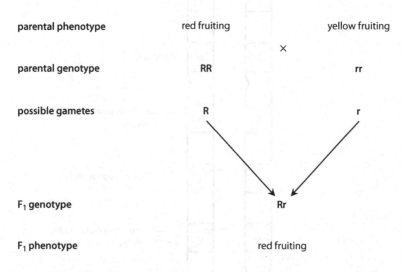

Figure 2.3 Producing the F_1 generation: a cross between red-fruiting and yellow-fruiting tomatoes.

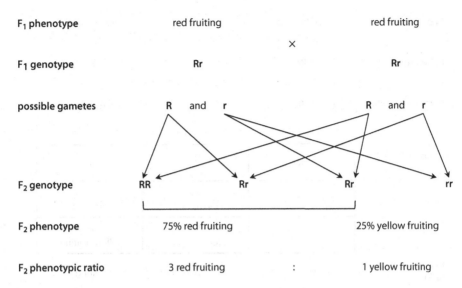

Figure 2.4 Producing the F_2 generation: a cross between two F_1 plants from Figure 2.3.

2.4 A Punnett square

Instead of using lines to represent possible fertilizations, many geneticists prefer to use a grid or **Punnett square,** named after the geneticist Reginald Punnett (1875–1967) who first used this method. By using a Punnett square (Figure 2.5), there is less chance of making a mistake (e.g. of forgetting a possible fertilization) than by drawing lines between gametes, as in Figure 2.4.

2.5 Mendel's First Law of Segregation

The results of crossing pure-breeding red-fruiting tomatoes and pure-breeding yellow-fruiting tomatoes closely mimics experiments performed by the "Father of Genetics," Gregor Mendel. Mendel's contribution to genetics is so important that the adjective "Mendelian" is often used to describe the kind of experiments that he originally carried out and the principles he formulated.

In his series of carefully planned experiments in his monastery garden at Brno, Gregor Mendel laid the foundations of the modern science of genetics. He worked with plants of the garden pea, *Pisum sativum,* and followed the inheritance of various pairs of contrasting characters (e.g. round or wrinkled seeds, white or purple flowers, green or yellow seeds, and long or short stems) through many generations. He carefully counted the numbers of individuals in each generation showing different alternative phenotypes. The validity of his findings is strengthened by the large numbers of plants with which he worked (often thousands).

F_1 phenotype	red fruiting		red fruiting
F_1 genotype	Rr	×	Rr
possible gametes	R and r		R and r

F_2 result

gametes	R	r
R	RR red fruiting	Rr red fruiting
r	Rr red fruiting	rr yellow fruiting

F_2 phenotypic ratio   3 red fruiting : 1 yellow fruiting

Figure 2.5 Using a Punnett square to predict the results of interbreeding F_1 plants.

Although not expressed in the genetic language that we use today, he drew the same conclusions (summarized in Table 2.1) from his experiments as discussed previously with respect to inheritance of fruit color in tomatoes. Mendel's conclusions led him to propose the first of two important laws, **The Law of Segregation**, which states that:

In the formation of gametes the paired hereditary determinates separate (segregate) in such a way that each gamete is equally likely to contain either member of a pair.

Mendel's "hereditary determinants" are what we now call alleles and, as discussed in Section 5.10, meiosis ensures that each gamete contains just one of a pair of alleles. Fertilization restores the normal paired situation.

TABLE 2.1 A summary of Mendel's conclusions

Traits of an organism are determined by particulate factors
Each parent has two of these particles
Pure-breeding strains contain a pair of identical particles
F_1 hybrids have two different particles
Only one of a pair of different particles are expressed in the F_1 hybrid
Particles are transmitted from parents to progeny through gametes
Each gamete contains just one particle
It is random which one of a pair of particles enters a gamete
Gametes, and so particles, randomly unite in a zygote

2.6   Predicting the outcome of crosses

By considering the pattern of inheritance of fruit color in tomatoes, certain fundamental rules of inheritance have been illustrated. A knowledge of these rules (e.g. a dominance relationship exists between alleles and that crosses involving two heterozygotes produce offspring in a 3 : 1 phenotypic ratio) enables us to make predictions about likely phenotypes and genotypes in future generations. This can be very helpful; for example, for prospective parents to know the likelihood of any children expressing a given disease.

It is important, however, to remember that genetic predictions are not absolute – they give us only the **chance** or **probability** of certain outcomes. Parents can know, for example, that they each have a particular genotype that indicates that they have a 1 in 4 chance of having a child with cystic fibrosis, but this outcome is not a certainty. It is essential to appreciate this probability or chance aspect to inheritance. A major part of transmission genetics involves making predictions about the chances of certain outcomes over others. Returning to the example this chapter has been using, when the F_1 red-fruiting tomatoes were crossed among each other, the outcome was presented as 3/4 red-fruiting and 1/4 yellow-fruiting plants in the F_2 generation. This was only a prediction in an ideal world. The observed numbers are more likely to only approximate to a 3 : 1 relationship.

Before introducing any further genetic principles it is necessary to stop and consider what is meant by "chance" and "probability." Why, practically, you could plant, say, 20 seeds from an F_1 cross and not obtain a single yellow-fruiting tomato plant, although theory predicts you would get five.

2.7   Chance and probability in genetics

When introducing ideas about probability it helps to use familiar situations; for example, when tossing a coin we all know that it is equally likely that heads or tails will be uppermost. The probability of throwing a head is thus described as 1 out of 2, 0.5, or 1/2. This is defined formally as:

$$\text{Probability of a particular outcome} = \frac{\text{number of ways of getting a particular outcome}}{\text{total number of possible outcomes}}$$

So, with our coin example:

- There is only *one* way of obtaining a head.
- There are *two* possible outcomes (heads and tails).
- Therefore, the probability of obtaining heads = 1/2.

Consider another, often cited example – the pack of playing cards. What would be the probability of obtaining a seven? The answer is 1/13, by the following reasoning:

- There are *four* ways of obtaining seven (four sevens in a pack).
- There are 52 possible outcomes (52 cards in a pack).
- Therefore, the probability of obtaining a seven = 4/52 = 1/13.

Referring back to our F_1 genetic cross (Figure 2.4), we could ask what is the probability of obtaining a yellow fruiter:

- There is *one* way of obtaining a yellow fruiter ($\mathbf{r} \times \mathbf{r}$).
- There are *four* possible outcomes ($\mathbf{R} \times \mathbf{R}$, $\mathbf{R} \times \mathbf{r}$, $\mathbf{r} \times \mathbf{R}$, and $\mathbf{r} \times \mathbf{r}$).
- Therefore, the probability of obtaining a yellow fruiter = 1/4.

One other idea should be appreciated here. When considering the number of possible outcomes, each is *equally likely*. You are equally likely to throw a head or a tail, or to pull any one of the 52 different cards from the pack. Each heterozygous F_1 red fruiter produces equal numbers of $\mathbf{R}$ and $\mathbf{r}$ gametes, and each is equally likely to be used at fertilization.

2.8   The importance of large numbers

Probability gives us theoretical expectations on the premise that all possible outcomes are equally likely. Consider the case of two parents who already have five daughters and desperately want a son. Surely the next conception will give them their longed-for son. After all, with six children they could have expected three daughters and three sons. But no – the sixth child is another daughter! Reality often deviates from expectation and so it is with genetic experiments. Consider again the F_1 red-fruiting heterozygote. There is no guarantee that it will produce exactly equal numbers of $\mathbf{R}$ and $\mathbf{r}$ gametes; there might be, for example, differences in gamete viability. Furthermore, random fertilization rarely uses each kind of gamete exactly equally.

To illustrate this latter idea, consider again the act of throwing a coin. The expected outcome is an equal number of heads or tails uppermost: a 1 : 1 ratio. Ten throws of the coin might produce a marked deviation from a 1 : 1 ratio. Yet after 1000 throws the ratio is likely to be very close to 1 : 1. Deviations from the expected are fewer the more times a probability event is performed – the *observed* result is more likely to equal the *expected*. These considerations stress the importance of using large numbers of individuals in genetic crosses, so that offspring ratios are meaningful and valid interpretations can be made. For example, if a cross yields a 3 : 1 ratio of two different phenotypes, we can accept this as a true reflection of the underlying genetics; in this case, that each parent was a heterozygote.

In conclusion:

- Offspring ratios are often enormously important in indicating the genetic basis of inheritance for a given trait.
- To ensure observed ratios are meaningful, large numbers of offspring must be produced (Mendel appreciated this; see Table 2.2).
- We can then match observed ratios to one of a set of expected ratios for different genetic situations.

The 3 : 1 phenotypic ratio is one of a number of useful genetic outcomes. This chapter has focused on this one ratio, important in the context of monohybrid crosses. The 1 : 1 ratio is another very useful monohybrid ratio, discussed below.

2.9   The test cross

As yellow fruit color in the tomato is recessive to red fruit color, we know immediately the genotype of any yellow-fruiting plant – it is homozygous recessive, **rr**. However, a red-fruiting plant could be homozygous or heterozygous dominant, **RR** or **Rr**. It is possible to distinguish between these alternatives for any red-fruiting plant by crossing the red fruiter with a yellow-fruiting plant. The phenotypes of the offspring indicate the genotype of the red parent. If the **test cross** yields approximately equal numbers of red- and yellow-fruiting plants (indeed any significant numbers of yellow fruiters) this indicates that the

TABLE 2.2 Results of Mendel's monohybrid experiments (indicating a clear 3 : 1 pattern!)

PARENTAL TRAITS	NUMBER OF F_2 PROGENY	F_2 RATIO
Round × wrinkled seeds	5474 round, 1850 wrinkled	2.96 : 1
Yellow × green seeds	6022 yellow, 2001 green	3.01 : 1
Inflated × constricted pods	882 inflated, 299 constricted	2.95 : 1
Green × yellow pods	428 green, 152 yellow	2.82 : 1
Purple × white flowers	705 purple, 224 white	3.15 : 1
Axial × terminal flowers	651 axial, 207 terminal	3.14 : 1
Long × short stems	787 long, 277 short	2.84 : 1

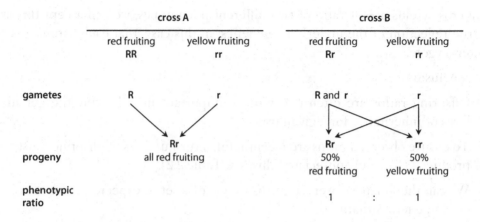

Figure 2.6 Results of a test cross if the red fruiter is homozygous (cross A) or if the red fruiter is heterozygous (cross B).

red-fruiting plant is a heterozygote. Otherwise, the conclusion is that the plant is homozygous (Figure 2.6). Thus, crossing an individual of dominant phenotype, but unknown genotype, with a recessive homozygote is a common and useful test of genotype. If a reasonable number of offspring, usually about 50%, show the recessive phenotype, then it can be concluded that the individual showing a dominant phenotype has a heterozygous genotype.

Summary

- Monohybrid crosses investigate the genetic basis of characters determined by a single gene.

- Genes can exist in alternative forms or alleles.

- Each individual possesses two similar or different alleles for a given trait.

- Alleles are passed on unchanged to the next generation.

- Alleles segregate into gametes at meiosis and come randomly together again when gametes unite at fertilization.

- The ratio of different types of progeny indicate the genotypes of parents:

  a 3 : 1 phenotypic ratio indicates heterozygous parents (**Aa** × **Aa**);

  a 1 : 1 phenotypic ratio indicates that one parent is heterozygous and the other homozygous recessive (**Aa** × **aa**).

Problems

1. In mice, black eye color is dominant to red. A pure-breeding black-eyed male was crossed with a female who was pure breeding for red eyes.

   (a) What phenotype could you expect in the F_1 generation?

   (b) If the eight F_1 mice were allowed to interbreed, what phenotypes would you expect among the F_2 generation and in what proportions?

2. If an F_2 mouse from Question 1 has black eyes, how can you decide whether it is homozygous or heterozygous?

3. Two short-haired guinea pigs were mated on three separate occasions and produced 21 short-haired and six long-haired offspring.

   (a) What is the genetic basis of hair length in guinea pigs?

   (b) What are the genotypes of the short-haired parents?

4. The author Ernest Hemingway was famous for his love of cats; as many as 60 lived with him at his house at Key West, Florida, USA. His cats were also famous – for having six toes! If a normal five-toed stray appeared and mated with one of these six-toed cats, the result was a litter consisting largely or entirely of six-toed cats! What does this suggest to you about the genetic basis of polydactyly (extra toes) in cats?

5. A variety of poppies that are pure breeding for spots at the base of their petals were crossed with another, non-spotted variety. The F_1 plants were allowed to self-fertilize, with a resulting F_2 generation consisting of 264 spotted and 84 unspotted poppies. What was the phenotype of the F_1 poppies?

6. Dimples are a dominant trait in humans. A man who is homozygous for dimples and a woman without dimples have children. What are their chances of having a child with dimples?

7. In the fruit fly, *Drosophila melanogaster*, gray body color (**G**) is dominant to black body color (**g**). A geneticist had three flies with gray bodies, designated P, Q, and R. He crossed P and Q, and obtained 109 gray-bodied flies. Q and R gave 80 gray-bodied and 28 black-bodied flies, while P and R gave 76 gray-bodied flies. What would be the expected genotypic and phenotypic ratios when flies P, Q, and R are crossed with black flies?

8. A tobacco grower crossed pure-breeding plants with large leaves with pure-breeding plants with small leaves. The F_1 plants all had large leaves. He allowed these F_1 plants to self-fertilize. The resulting seed when planted, produced 640 plants. How many of these would you expect to be large leaved and how many small leaved?

9. In humans, brown eyes is dominant to blue eyes and identical twins occur approximately once in every 300 births. What is the probability of a blue-eyed couple having brown-eyed identical twin boys as their first children?

10. Pecan shells are sometimes thin and liable to shatter during harvest. A farmer wanted to make sure that he would not lose any of his valuable crop through this undesirable trait. Thus, he investigated the genetic basis of this trait by taking pollen from two different trees (A and B), both of which produced thick-shelled pecans, and using it to fertilize flowers on a tree that produced thin-shelled pecans. Pollen from tree A produced 56 plants, of which 29 eventually produced thick-shelled nuts, while pollen from tree B resulted in plants that all produced thick-shelled pecans. What is the mechanism of inheritance of thick and thin pecan shells?

11. In cattle, to be polled (hornless) is dominant to horned.

(a) What are the genotypes of polled parents that produce a calf which subsequently grows horns?

(b) What is the probability that any subsequent calves born to these parents are (i) polled; (ii) grow horns; (iii) produce all polled offspring when they mature, regardless of the genotype of their mates?

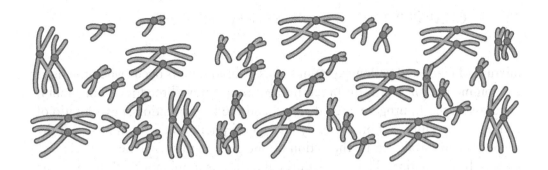

Extensions to Monohybrid Inheritance

The analysis of phenotypic ratios among the progeny of a mating provides important clues to the genetic basis of different traits. Chapter 2 introduced this important approach with respect to monohybrid crosses (i.e. to recognizing when a character was determined by one gene with two alleles).

This chapter introduces further ideas about monohybrid inheritance, in particular:

- Situations where there are more than two alleles possible for a particular gene
- Some differing dominance relationships between alleles
- How the rules are applied to human inheritance

3.1 Multiple allelism

As anyone knows who has kept rabbits as pets, they can have a wide variety of different coat colors. There is, for example, the diffuse gray/brown color of wild rabbits, known as agouti. This coloration is the result of each hair having a mixture of black and yellow pigment bands. Albino rabbits occur when the hairs lack pigmentation. If a pure-breeding agouti and pure-breeding albino rabbit are crossed, the F_1 progeny are all agouti and the F_2 generation shows a ratio of 3 : 1 agouti to albino rabbits. Clearly, the agouti phenotype is dominant and the two colors are produced by the action of one gene with two alleles – a "classic" monohybrid situation. That the agouti coloration is dominant to albino may be expected as the variously colored hairs of agouti rabbits obviously produce a better camouflaged rabbit than a brilliant white rabbit.

Albino rabbits are, however, popular as pets, as are white rabbits with black feet, ears, nose, and tail (Himalayans), and the delicate silver-gray chinchillas. If one gene determines coat color, how can the normal monohybrid inheritance rules account for these additional variants? This range of rabbit fur colors introduces a new genetic idea – that a gene may have more than two alleles. It can show **multiple allelism**. Although each individual can have no more than two alleles of one gene in its cells, many more can be present among the different members of a population. In fact the main gene determining fur color in rabbits has four alleles – agouti, chinchilla, Himalayan, and albino, with dominance in this descending order.

In other words, agouti pigmentation pattern is dominant to the other three pigment types and albino is recessive. Chinchilla, however, is dominant to Himalayan. With these allele relationships, four phenotypes can be produced by 10 different genotypes, shown in Table 3.1, which presents one of a variety of different ways of representing the alleles of a multiple allele series.

TABLE 3.1 Genetic control of rabbit coat color (allele symbols: C = agouti, C^{ch} = chinchilla, C^h = Himalayan, and C^a = albino)

GENOTYPE OF RABBIT	PHENOTYPE OF RABBIT
CC	Agouti
CC^{ch}	Agouti
CC^h	Agouti
CC^a	Agouti
$C^{ch}C^{ch}$	Chinchilla
$C^{ch}C^h$	Chinchilla
$C^{ch}C^a$	Chinchilla
C^hC^h	Himalayan
C^hC^a	Himalayan
C^aC^a	Albino

gametes	C^{ch}	C^a
C^h	$C^{ch}C^h$ chinchilla	C^hC^a Himalayan
C^a	$C^{ch}C^a$ chinchilla	C^aC^a albino

parental cross Himalayan C^hC^a × chinchilla $C^{ch}C^a$

results

phenotypic ratio   2 chinchilla : 1 Himalayan : 1 albino

Figure 3.1 The results of a mating between a Himalayan and a chinchilla rabbit.

If the dominant/recessive relationships between the different alleles in a multiple allele series are known, then any cross can be considered in the "normal" Mendelian way, as discussed in Chapter 2. We can, for example, explain the appearance of albino and Himalayan rabbits from a mating between a chinchilla doe and a Himalayan buck (i.e. both the doe and the buck must have been heterozygotes, carrying an albino allele) (Figure 3.1).

Many genes show multiple allelism; another example is the gene determining markings in tabby cats (Figure 3.2). Three alleles determine the different tabby types. The main dominant allele, **T**, produces the characteristic mackerel or striped tabby. T^a is responsible for the Abyssinian pattern, where tabby markings are restricted to the head and face. If neither of these two alleles is present, t^b will produce a blotched tabby coat pattern. In these two examples, rabbit fur color and tabby cat pattern, there are only a few alleles possible at the relevant loci – four and three alleles, respectively. There is, however, no limit to the number of alleles that can exist at a given locus. The gene coding for a dehydrogenase enzyme in the much-studied fruit fly, *Drosophila melanogaster*, possesses at least 32 alleles. Populations of the red clover, *Trifolium pratense*, can carry dozens of different alleles at a self-sterility locus (see Box 3.1 for further details), as can the major histocompatibility genes (MHC) of vertebrates that code for proteins of the immune system. Hundreds of alleles are currently known at one human immunoprotein locus, *DRB1*!

So why do some genes possess just two alleles and others 20? The existence of many alleles at a locus may be regarded as evolutionarily advantageous. Diversity is evolution's best strategy. Many alleles result in multiple phenotypes and maximize the chance of at least some individuals possessing features suitable for a given environment, thus enabling their survival and ultimately that of the species. In the self-sterility example (Box 3.1), the advantage is the prevention of inbreeding – a pollen tube will not grow through style tissue of a plant whose genotype includes the same self-sterility allele as the male tissue. Inbreeding

Figure 3.2 A striped tabby cat.

BOX 3.1 SELF-STERILITY IN PLANTS

The flowers of many plants are hermaphrodite (i.e. they contain both male and female reproductive organs). Various mechanisms exist to prevent self-fertilization. One strategy involves **self-sterility genes**. These genes, generally one major locus per species, possess multiple alleles. A pollen tube will not successfully grow down a style towards the egg in the ovary if the pollen carries either of the same self-sterility alleles as the female tissue. The figure shows the growth of pollen tubes through female tissue of different genotypes (pollen produced by a plant of genotype S_1S_8).

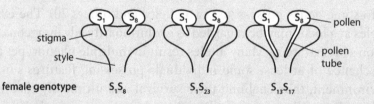

Box 3.1 Figure 1 Growth of pollen tubes through female tissue of different genotypes: pollen produced by a plant of genotype S_1S_8.

increases the incidence of homozygous recessive genotypes, many of which can be detrimental to an individual's survival chances. As a result, individuals die, removing valuable variants from a population's gene pool. A population's number may thus decrease, which can further compound the negative situation, as survivors are likely to be genetically more similar.

3.2 Incomplete dominance

Section 3.1 described the situation in which many allelic forms of a gene can exist within a population. These multiple alleles obey the same rules of expression as just two alleles, with clear dominance relationships between them. Complete dominant/recessive relationships do not, however, always exist. There are many examples of this **incomplete dominance** in plants. Flower color in the snapdragon, *Antirrhinum majus* (Figure 3.3), is determined by a single pair of alleles – one promoting red petals and the other promoting white petals. Chapter 2 established rules that should enable us to predict the outcome of crossing a pure-breeding red-flowering snapdragon with a pure-breeding white-flowering plant. One color would be expected to be dominant – presumably the stronger one,

Figure 3.3 Snapdragon flowers.

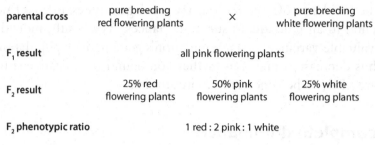

Figure 3.4 Investigating the inheritance of snapdragon flower color.

red. Thus, it would be reasonable to assume that all the F_1 plants would produce red flowers. But no! The flowers of the F_1 generation are all pink – neither color dominates. Instead, the red and white traits have apparently blended to give an F_1 generation of intermediate color to either parent. What happens therefore when these pink F_1 plants are crossed among themselves? The answer is that red-, pink-, and white-flowering plants are all observed in the F_2 generation, and in distinctive proportions: 1/4 are red, 1/4 are white, and 1/2 are pink-flowering (Figure 3.4). So, how can we explain these observations?

The fact that red and white flowers appear again in the F_2 generation shows that the color blending is at the phenotypic level. The red and white alleles have remained as discrete functional units when together in the F_1 plants so that they can segregate at gamete formation and recombine at fertilization to give all possible colored plants in the F_2 generation. Another genetic phenomenon is illustrated here – **incomplete dominance**. This occurs when neither of a pair of alleles shows dominant expression to the other. Thus, the heterozygote displays a distinctive phenotype to either of the two homozygotes. Indeed, the heterozygous phenotype is often intermediate in its expression to that of both homozygotes, as in this example of snapdragon flower color – pink is intermediate in intensity between red and white.

3.3   Representing crosses involving incomplete dominance

When displaying genetic crosses it is conventional to use an upper-case letter for a dominant allele and a lower-case letter for the corresponding recessive allele (Section 2.3). When there is no dominance relationship between two alleles, they must be represented differently. Generally a different capital letter is used for each allele. So, in the present example of red- and white-flowering snapdragons, **R** represents the red allele and **W** represents the white allele. The monohybrid cross could then be represented as shown in Figure 3.5.

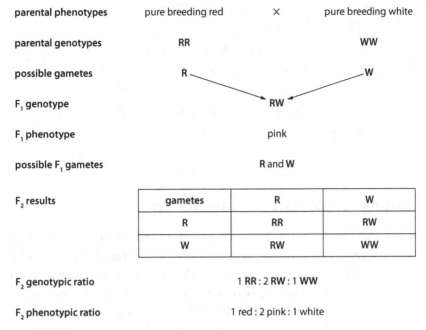

Figure 3.5 Representing the inheritance of snapdragon flower color.

Another informative Mendelian ratio is illustrated. If a cross produces three phenotypes in a 1 : 2 : 1 ratio, then a monohybrid situation involving two incompletely dominant alleles should be suspected, especially if the phenotype of the larger class is intermediate in expression between the other two phenotypes.

3.4   Explaining incomplete dominance

An explanation for incomplete dominance can often be found by considering the nature of the gene product. In such cases, only one of the two alleles produces a functional product. This means that the cells of one homozygote produce two units of working product, those of the heterozygote produce one unit, and those of the other homozygote produce none. Consider this interpretation within the context of the various petal colors of snapdragons. Red pigment is formed by a complex sequence of enzyme-controlled reactions. The **R** allele codes for a crucial enzyme in this pathway, while the **W** allele fails to produce an active enzyme. Thus, the **RR** homozygote produces sufficient enzyme for a lot of red pigment to be produced from a white precursor pigment. In heterozygotes, **RW**, only one allele is producing active enzyme, so only some of the white precursor pigment is changed. Both the unchanged white precursor and red pigment are present in the petals of these plants. As a result, the petals appear pink. Any plant of **WW** genotype is, of course, white. This situation is summarized in Table 3.2.

TABLE 3.2 The relationship between genotype, enzymes, and pigment in the determination of snapdragon petal color

GENOTYPE	STATE OF ENZYME	PIGMENT PRESENT	PHENOTYPE
RR	100% active	Red	Red
RW	50% active/50% inactive	Red and white	Pink
WW	100% inactive	White	White

3.5   Redefining dominance relationships

It is common for the heterozygote to show an intermediate phenotype to the two homozygotes when color phenotypes are being considered. Various other examples are given in the problems at the end of the chapter. Frequently, the different colors can be related to varying levels of functional product produced by two alleles, as in the snapdragon example. Indeed, in recent years, as analysis of an increasing number of phenotypes at the biochemical level has become possible, a redefinition of dominance relationships between pairs of alleles has been necessary.

Complete dominance of allele **A** over allele **a** means that the genotypes **AA** and **Aa** are indistinguishable phenotypically. However, it is becoming clear that the ability to distinguish the homozygous and heterozygous dominant genotypes is often a matter of how the phenotype is examined. Increasingly sophisticated biochemical tests are making it possible to distinguish phenotypically between the two dominant genotypes. An interesting illustration of this recent phenomenon is provided by a re-examination of Mendel's round and wrinkled peas. When considered at the gross morphological level, seeds are clearly either round or wrinkled in form, with roundness dominant to wrinkledness. However, when seeds are examined microscopically or biochemically, a different and less clearly defined picture emerges (Box 3.2). It is, therefore, perhaps a good thing that neither sophisticated microscopes nor biochemical techniques were available to Gregor Mendel when he was performing his momentous experiments!

Observations such as those with the garden pea necessitate a redefinition of dominance. Dominance should be seen as a property of a pair of alleles in relation to the particular attribute of the phenotype being examined. Phenotypes may have many different physical and biochemical attributes; complete dominance may be observed for some of these and not others. This blurring of dominance relationship between alleles may at first seem confusing. It does not, however, detract from the useful conclusions that can be drawn from analyzing the phenotypic ratios among the progeny of crosses. A phenotypic ratio of either 1 : 2 : 1

BOX 3.2 MENDEL'S ROUND AND WRINKLED PEAS

In 1991, a team of British geneticists identified the pea shape gene, and worked out the physiological basis of roundness and wrinkledness. Starch can exist in two different forms – unbranched (amylose) and highly branched (amylopectin). The pea shape gene encodes an enzyme known as **SBE1** (starch branching enzyme 1), which catalyzes the conversion of amylose to amylopectin. The dominant allele **R** causes the formation of an active SBE1 enzyme.

As a result, the seed of an **RR** homozygote is full of the branched amylopectin, which enables the pea to maintain a rounded shape and to shrink uniformly as it ripens. In contrast, the product of the **r** allele is a non-effective enzyme. The seed of an **rr** homozygote instead contains amylose. It is irregular in shape and shrinks unevenly as it ripens. The result is a wrinkled seed. Although the heterozygote, **Rr**, has reduced levels of the active SBE1 enzyme, it is sufficient to convert most of the amylose to amylopectin; the amylopectin content is high enough to result in uniform shrinking and so a round seed.

 Thus, if pea shape is assessed morphologically, roundness is completely dominant to wrinkledness. If levels of the SBE1 enzyme produced by the pea shape gene are assayed, the heterozygote has reduced enzyme levels and is incompletely dominant.

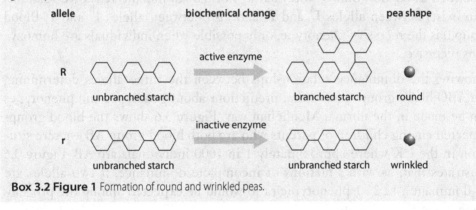

Box 3.2 Figure 1 Formation of round and wrinkled peas.

or 3 : 1 among offspring indicates the same thing – that one gene with two alleles is determining a given trait and that a cross has been made between two heterozygotes. The different ratios then tell us whether or not the heterozygote can be distinguished phenotypically from the dominant homozygote (i.e. whether the two alleles show complete or incomplete dominance).

3.6 Co-dominance

Rather than a heterozygote showing an intermediate phenotype to either homozygote, it is possible for its expression to be additive – both alleles make an equal contribution to the phenotype so that the heterozygote shows a distinctive phenotype which combines features of both homozygous traits. This situation is described as **co-dominance** and is most commonly observed with biochemical traits. A good, and often quoted, example of co-dominance is the relationship between two of the three alleles that determine the human ABO blood groups.

The ABO blood groups result from a variation in two different oligosaccharides, which we designate **antigen A** and **antigen B**, present on the surface of red blood cells. The ABO blood group gene encodes an enzyme that adds a final sugar to a precursor five-sugar oligosaccharide, known as the **H antigen**. Antigen A is synthesized by an enzyme encoded by allele I^A and antigen B is synthesized by an enzyme encoded by allele I^B. A third allele, I^i, produces no functional enzyme. Thus, the gene determining the ABO blood groups also shows multiple allelism (Section 3.1). Table 3.3 shows the relationship between the various genotypes and phenotypes.

The AB blood group is produced in individuals of heterozygous genotype $I^A I^B$. Each allele produces a functional (and different) sugar-adding enzyme, and so two different antigens, A and B, are produced. In the $I^A I^B$ heterozygote, therefore, both alleles are contributing equally to the phenotype. These alleles are described as **co-dominant**. Note that a "normal" dominant/recessive relationship exists between alleles I^A and I^i, and also between alleles I^B and I^i. Blood group 0 is the recessive phenotype, only possible when individuals are homozygous recessive.

Knowing the dominance relationships between the three alleles determining the ABO blood groups means that predictions about genotypes and phenotypes can be made in the normal Mendelian way. Figure 3.6 shows the blood groups expected among children of parents who are both blood group AB – a rare situation in the UK where approximately 1 in 1000 individuals are AB. Figure 3.6 illustrates that, as with situations of incomplete dominance, if two alleles are co-dominant a 1 : 2 : 1 phenotypic ratio would be expected among the progeny when heterozygotes mate.

TABLE 3.3 The genetic control of human ABO blood groups

GENOTYPE	ANTIGEN(S) PRESENT	BLOOD GROUP
$I^A I^A$	A	A
$I^A I^i$	A	A
$I^B I^B$	B	B
$I^B I^i$	B	B
$I^A I^B$	A and B	AB
$I^i I^i$	H	O

parental cross	AB × AB	
	I^AI^B I^AI^B	
possible gametes	I^A I^B I^A I^B	

gametes	I^A	I^B
I^A	I^AI^A blood group A	I^AI^B blood group AB
I^B	I^AI^B blood group AB	I^BI^B blood group B

phenotypic ratio
of blood groups
among children   1 A : 2 AB : 1 B

Figure 3.6 Predicting ABO blood groups.

3.7 Lethal alleles

There are certain genes whose products are essential for life. Any allele that codes for a faulty product will therefore be lethal in the homozygous state. An individual heterozygous for a lethal allele generally shows an affected, but less severe, phenotype. One striking example of the effect of a lethal allele is the cat with no tail – the Manx cat (Figure 3.7). All Manx cats are heterozygotes. In the homozygous state, the tail-less allele is believed to cause such extreme spinal defects as to be non-conducive to life; homozygous embryos are reabsorbed by the mother cat.

Figure 3.7 A Manx cat. Courtesy of S. J. Pyrotechnic under CC BY-SA 2.0 license.

| parental cross | tail-less Manx | × | tail-less Manx |
| | Mm | | Mm |

| possible gametes | M   m | | M   m |

gametes	M	m
M	MM lethal	Mm Manx
m	Mm Manx	mm tailed

phenotypic ratio among kittens   2 tail-less Manx : 1 normal tailed

Figure 3.8 The genetic consequences of mating Manx cats.

As Manx cats are heterozygotes, we might initially expect a 3 : 1 phenotypic ratio amongst the offspring of a mating between two Manx cats. Breeders of Manx cats, however, report a different outcome – smaller than average litter sizes with a 2 : 1 phenotypic ratio among the kittens (i.e. twice as many tail-less to tailed kittens). Figure 3.8 explains this result. One resulting genotype is lethal and so an unusual 2 : 1 phenotypic ratio is observed among the surviving progeny.

This example of inheritance of tail length in cats illustrates another useful Mendelian ratio. If two individuals of the same phenotype are crossed and two phenotypes are observed in a 2 : 1 ratio among the offspring, we can suspect a monohybrid situation with one of the two alleles being lethal in the homozygous state. But, is the lethal allele dominant or recessive?

3.8   Are lethal alleles recessive or dominant?

A paradox exists when considering lethal alleles. As the lethality is only expressed when the allele is homozygous, it seems to be behaving as a recessive allele. Yet when considered from the viewpoint of the allele's effect on the living organism, it is behaving as a dominant allele, as in the example of the absence of tails in Manx cats – the heterozygotes are tail-less. This apparently dual property of an allele illustrates an important aspect of all alleles. Dominance and recessiveness are not intrinsic properties of the alleles themselves, but depend upon the context. Thus, the **M** allele, in the example of the Manx cat, is dominant with respect to its effect on tail length, but recessive when viability is considered. Such alleles, which are lethal when homozygous but not when heterozygous, are often called **recessive lethal alleles**. They are recognized by the modification to the expected phenotypic ratio among the progeny of two heterozygotes – a ratio of 2 : 1 dominant to recessive individuals, instead of the expected 3 : 1 ratio for cases

TABLE 3.4 Phenotype ratios among the progeny of heterozygotes of monohybrid crosses involving one gene with two alleles

	PHENOTYPE		
	1	2	3
One allele completely dominant	3	1	
Two alleles incompletely or co-dominant	1	2	1
One allele lethal as homozygote	2	1	

of complete dominance or 1 : 2 : 1 for incomplete or co-dominance. These ideas are summarized in Table 3.4.

Occasional examples of a **dominant lethal allele** have been identified. Dominant lethal alleles exert their effect in both heterozygotes and homozygotes. They can only be recognized and studied genetically if death occurs after the individual has reached reproductive age. One example of a dominant lethal allele is that which causes the human condition referred to as Huntingdon's disease, which is discussed in Section 3.9.

3.9 Human pedigrees

The laws of genetics are universal. Inheritance operates in the same basic way in all organisms. However, when analyzing inheritance patterns in humans there is one fundamental difference – we cannot do it by examining the results of carefully orchestrated crosses! Furthermore, human families tend to be small and so it is impossible to identify ratios of phenotypic expression. Thus, a different strategy is adopted. A family tree or **pedigree** is constructed. The segregation pattern of a trait through several generations of related individuals often yields clues to its mode of inheritance. Figure 3.9 shows a pedigree for a human family in which the neurodegenerative Huntingdon's disease is found. The pedigree in Figure 3.9 shows that the trait:

- Affects both sexes
- Appears in each generation within an affected family
- Affects about half of all the children in an affected family

These are characteristic segregation features for a condition caused by the presence of a dominant allele at a single locus. In the pedigree of Figure 3.9, affected individuals are heterozygotes, **Hh**, and non-affected people are homozygous normal, **hh**. For individuals with one dominant allele, symptoms (involuntary movements and progressive central nervous system degeneration) do not

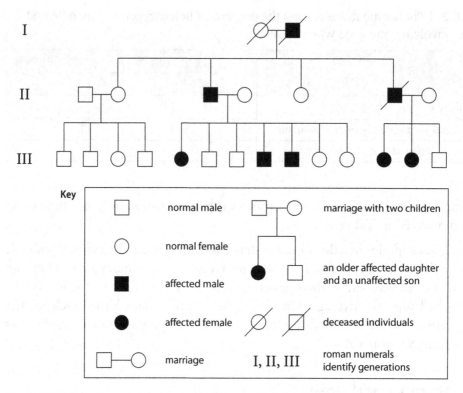

Figure 3.9 Pedigree illustrating the inheritance of Huntingdon's disease.

generally begin to show until they are in their 30s, with death following 15–20 years later. Although the disease allele is common within the family represented in Figure 3.9, it is important to remember that this dominant allele causing Huntington's disease is very rare – only around 1 in 10,000 individuals possess the allele.

Figure 3.10 shows a typical pedigree segregation pattern for a recessive trait; in this case, cystic fibrosis – a disease that causes dysfunction of the lungs, pancreas, and digestive system. The condition:

- Affects both sexes

- Is not observed in each generation

- Has a low frequency of expression (especially when compared with the dominant trait in Figure 3.9)

The pedigree in Figure 3.10 illustrates another characteristic feature of recessive traits, particularly rare ones – parents of affected individuals are often related. The parents of IV-1 to IV-3 are first cousins. They are also heterozygotes that each passed their recessive allele onto IV-3. The recessive cystic fibrosis allele

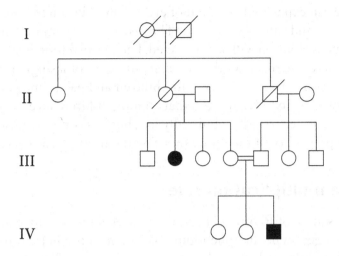

Figure 3.10 Pedigree illustrating the inheritance of cystic fibrosis.

(causing a defect in a protein controlling transmembrane chloride transport) must have been "silently" passed from parent to children for many generations. It is only when two heterozygotes, or **carriers**, have children together that there is the chance of a homozygous expressing state. Each child of these carrier parents (III-4 and III-5) had a 1 in 4 chance of expressing the recessive condition. This can be demonstrated by drawing out the cross as we have been doing before (Figure 3.11).

The Punnett square in Figure 3.11 shows the possible genotypes and phenotypes when two heterozygous individuals have children. They have a 1 in 4 chance of conceiving a child expressing the recessive phenotype. It does not, however,

parents	CFcf	×	CFcf

possible gametes   CF   cf   CF   cf

gametes	CF	cf
CF	CFCF normal	CFcf normal
cf	CFcf normal	cfcf cystic fibrosis

phenotypic ratio
among children   3 normal : 1 cystic fibrosis

Figure 3.11 The probability of heterozygous parents having children expressing cystic fibrosis. **CF** = normal transporter allele and **cf** = allele producing the faulty protein.

mean that we can expect only one out of every four children to have cystic fibrosis. If their first child suffers from cystic fibrosis, it is no guarantee that a couple's three subsequent children will be unaffected. Each fertilization is an independent event. A Punnett square shows the possible outcomes of a single fertilization. It is possible for two carriers to have four children and each time gametes carrying the recessive allele to fertilize, so that all four children have the disease. The probability of this occurring is admittedly low, but it is certainly possible. We can calculate the possibility of this situation using the **multiplication rule**.

3.10   The multiplication rule

The Punnett square of Figure 3.11 shows the probability of any conceived child having a given genotype and phenotype. We want to predict the probability of four successive children having the same genotype and phenotype (i.e. to be a recessive homozygote and suffer from cystic fibrosis).

Each fertilization is an independent event with a given probability. The probability that the same event occurs in successive fertilizations is calculated by **multiplying together the probability of each independent event**. Thus, in the example under discussion:

- The probability that a child expresses cystic fibrosis is $1/4$.

- The probability that four successive children express cystic fibrosis is $1/4 \times 1/4 \times 1/4 \times 1/4 = 1/256$.

The multiplication rule enables us to define any sequence of events and to work out their probability. We might, for example, be interested to know what is the probability of a first child suffering from cystic fibrosis and a second being disease-free. Referring again to the Punnett square of Figure 3.11:

- The probability of a child expressing cystic fibrosis is $1/4$.

- The probability of a child not expressing cystic fibrosis is $3/4$.

- The probability of the first child expressing cystic fibrosis and the second child not expressing cystic fibrosis is $1/4 \times 3/4 = 3/16$.

We could have asked a more general question. A couple know that they are both carriers for the cystic fibrosis allele. They want to know the probability, if they have two children, of one having the disease. This could be achieved in one of two ways:

1. First child has cystic fibrosis; second child disease-free

2. First child disease-free: second child has cystic fibrosis

The probability of each of the above two events is:

1. $1/4 \times 3/4 = 3/16$
2. $3/4 \times 1/4 = 3/16$

As either of the two scenarios is possible, they both have to be taken into consideration when calculating a final probability, so now the **addition rule** applies:

Probability of one of two children having cystic fibrosis

= probability of pattern 1 + probability of pattern 2

= $3/16 + 3/16 = 6/16$, or $3/8$

Thus, the answer to the couple's question about the probability of one of two children having cystic fibrosis is $3/8$.

It needs practice with a variety of different situations to confidently know when to multiply and when to add probabilities. Some examples are included in the problems at the end of the chapter.

Summary

- Many alleles can exist at each gene locus. One individual, however, can possess only two alleles, so the inheritance of multiple alleles conforms to Mendel's Law of Segregation.

- Alleles at a locus may not show a clear-cut dominant/recessive relationship. Dominance relationships between alleles often need to be carefully defined in terms of the particular aspect of the phenotype being examined (e.g. physical, behavioral, or biochemical).

- Only if there is no detectable difference between a homozygote and a heterozygote can an allele be described as completely dominant. Otherwise, we recognize incomplete dominance and co-dominance.

- A 1 : 2 : 1 phenotypic and genotypic ratio in a cross between two heterozygotes indicates alleles are incompletely dominant or co-dominant.

- A 2 : 1 phenotypic ratio between two heterozygotes indicates an allele is lethal as a homozygote.

- Distinctive segregation patterns in human pedigrees can indicate whether a specific condition results from a dominant or recessive allele at a given locus.

Problems

1. In the four o'clock plant, the allele for red flowers is incompletely dominant over the allele for white flowers, so the heterozygotes are pink. What ratio of flower colors would you expect among the offspring of the following crosses?

   (a) Pink-flowering × pink-flowering plants.

   (b) White-flowering × pink-flowering plants.

   (c) White-flowering × red-flowering plants.

   (d) White-flowering × white-flowering plants.

2. Henrik is blood group B, like his mother. His older sister is group A and his younger brother is group O. What are the genotypes of his parents?

3. Three alleles at a single locus determine whether guinea pig coat color is yellow, cream, or white. One homozygote, $C^Y C^Y$, is yellow, the other homozygote is white, $C^W C^W$, while the heterozygote, $C^W C^Y$, is cream. If two cream-colored individuals are mated, what phenotypes will be observed among the progeny and in what ratios?

4. The shape of radishes may be long, $S^L S^L$, round, $S^R S^R$, or oval, $S^L S^R$. If plants producing long radishes are crossed with some producing round radishes, and the resulting F_1 plants selfed, will any oval-producing plants be present in the F_2 generation?

5. In 1943, actress Joan Barry sued Charlie Chaplin for support of a child that she claimed he had fathered. She was blood group A, Chaplin group O, and the child group B. Chaplin was deemed the father and ordered by the court to pay maintenance for the child. Do you agree with the court's decision?

6. In chickens, the dominant allele (C) produces 'creepers' (short-legged chickens) when it is heterozygous, but when homozygous the offspring never hatch. A creeper cockerel and hen are mated, and produce 21 viable offspring. How many do you expect to have normal legs?

7. It was wondered whether migratory behavior had a genetic basis. A system was developed for studying migration in the laboratory. It was observed that birds that migrated became much more energetic at the appropriate times of the year. Sensors were therefore fitted to their perches and the number of jumping movements, measured electronically, correlated with migratory behavior. Crosses were made between birds from two populations of the blackcap warbler: a non-migratory population that lived all year in Africa, and another population that spent the summers in Germany and the winters

in Africa. All the resulting F_1 birds showed jumping movements intermediate in number between the two parents. Suggest:

(a) What the genetic basis of migratory behavior is.

(b) What the result would be of crossing the F_1 birds with birds from the non-migratory African population.

8. One gene with five alleles controls the patterning on lentil seeds. Two of the five alleles produce two different marbled patterns (L^{M1} and L^{M2}). The other three alleles result in spotted (L^S), dotted (L^D), or clear (L^C) seeds. The dominance relationship is in the order in which these alleles have been presented (i.e. L^{M1} is dominant to the other four, then L^{M2}, L^S and L^D, and finally L^C is recessive to all others). Thus, what would be the expected phenotypic ratios among the plants of the following crosses?

(a) Marbled-1 ($L^{M1}L^{M1}$) × spotted ($L^S L^S$).

(b) Marbled-2 ($L^{M2}L^C$) × dotted ($L^D L^C$).

(c) Marbled-1 ($L^{M1}L^D$) × marbled-2 ($L^{M2}L^D$).

9. In foxes, two alleles exist at a particular locus, **P** and **p**. The homozygote, **PP**, results in lethality, the heterozygote produces a platinum coat, while the homozygote, **pp**, produces a silver coat. What phenotypes, and in what ratios, are obtained when platinum foxes breed?

10. Horses can be cremello (a light cream color), chestnut (a deep brown), or palomino (golden with a white tail and mane). Cremello and chestnut horses always breed true, but palominos never do. From the results below, deduce the mode of inheritance of these colors and complete the table below, showing the results of matings between cremello, chestnut, and palomino horses.

Parents	F_1 progeny	Parental genotypes
Cremello × palomino	50% cremello; 50% palomino	
Chestnut × palomino	50% chestnut; 50% palomino	
Palomino × palomino	25% cremello; 50% palomino; 25% palomino	

11. Below is a pedigree for an inherited form of deafness (affected individuals are represented by a solid symbol).

(a) Is the mode of inheritance of this trait recessive or dominant?

(b) Choose suitable systems, and give the genotypes of individuals (i) 11-1, (ii) 11-3, (iii) 111-2, and (iv) IV-3.

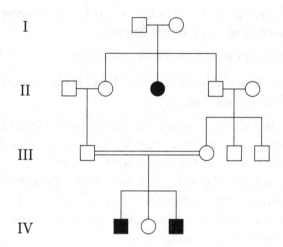

12. In pigs, the normal cloven-footed condition is dominant to mule footed. A true-breeding cloven-footed pig is crossed to a mule-footed pig. The resulting F_1 pigs are self-fertilized. What are the following probabilities for the F_2 generations:

(a) That the first pig examined is mule footed?

(b) That the first pig examined is mule or cloven footed?

(c) That the first three pigs examined are cloven footed?

13. In fowl, the alleles for black and white feathers are co-dominant. The heterozygote, therefore, produces both black and white feathers, and the resulting color is 'blue' or 'erminette'. If an erminette cockerel mates with an erminette hen, what is the probability that from a clutch of four eggs:

(a) All will hatch into erminette fowls?

(b) All will hatch into white fowls?

(c) The first egg will hatch into a white fowl and the other three eggs produce black fowls?

14. Schilder's disease is a progressive degeneration of the central nervous system in humans. It is inherited as a recessive trait. A couple lose their first child to the disease. They would very much like a family of two children. What is the probability that two further conceptions will produce disease-free children?

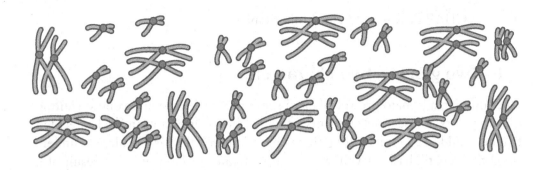

Dihybrid Inheritance

In the previous two chapters we have been considering traits determined by allelic variation at a single locus. Such monohybrid traits are, however, in the minority. The vast majority of traits in all organisms arise from the action of two or more, often many, genes. In addition, interactions between genes and the environment can be important in determining the final phenotype. The outcome of controlled breeding experiments and analysis of ratios among progeny can still provide important clues to the genetic basis of such traits. This chapter focuses on various ways in which two genes can interact to determine a single trait. Identifying these situations requires an understanding of the principles of **dihybrid inheritance** – crosses that simultaneously consider the inheritance of alleles at two separate loci. Gregor Mendel first performed these kinds of crosses; he was interested in examining patterns of inheritance of two different traits, each controlled by a single gene. He discovered that the F_2 generation showed some characteristic ratios, which remain extremely useful in flagging up the involvement of two genes in the expression of a trait.

This chapter introduces the expected patterns of phenotypic expression in the F_1 and F_2 generations when:

- Two genes control the expression of two separate characteristics.
- Two genes control the expression of a single trait.

4.1   Two genes – two characters

One summer, a tomato grower planted some purple-leaved and yellow-fruiting tomato plants. These unusual plants attracted a lot of attention and so he decided that he would like to start selling them on a commercial basis. He already understood the genetic basis of yellow-fruiting tomatoes – that it was the result of a recessive allele at a single locus (see Chapter 2). He was, however, unaware of the genetic basis of leaf color and how or whether it was possible to consistently breed his desired combination of characteristics. The tomato grower was in luck. Mendelian principles could be applied to ensure a constant supply of the desired phenotype.

To understand the principles involved in selectively breeding plants with the desired two characters, let us return to the standard procedure used in Chapter 2 when establishing the principles of monohybrid inheritance (i.e. that of crossing pure-breeding parents to achieve an F_1 generation that is then crossed among itself to produce the F_2 generation). In this current example, two characteristics are under consideration:

- Fruit color: either red or yellow

- Leaf color: either green or purple

Suppose pure-breeding yellow-fruiting, purple-leaved plants are crossed with pure-breeding red-fruiting, green-leaved plants. The resulting phenotypes of the F_1 and F_2 plants are shown in Figure 4.1.

As with monohybrid examples, a striking pattern of inheritance can be seen – certain phenotypes have disappeared in the F_1 generation, but reappeared again in the F_2 generation. Indeed, four different phenotypes are observed in the F_2 generation, representing all possible combinations of the individual characters. If this experiment is repeated enough times, a consistent relationship between the different F_2 phenotypes emerges – the four different phenotypes of red-fruiting,

parental cross	pure breeding yellow fruiting, purple leaved		×	pure breeding red fruiting, green leaved	
F_1 result			red fruiting, green leaved		
F_2 result	red fruiting, green leaved	red fruiting, purple leaved		yellow fruiting, green leaved	yellow fruiting, purple leaved
F_2 ratio	9	:	3	:	3	:	1

Figure 4.1 Investigating the inheritance of tomato fruit and leaf color.

green-leaved, red-fruiting, purple-leaved, yellow-fruiting, green-leaved, and yellow-fruiting, purple-leaved plants occur in a recognizable ratio of 9 : 3 : 3 : 1.

Taking our cues from the monohybrid work, we can conclude that the phenotypes expressed in the F_1 generation are the dominant phenotypes (here, red fruiting and green leaved). We can therefore rewrite this current cross, using symbols for the different alleles:

- Let **R** = red fruiting
- Let **r** = yellow fruiting
- Let **G** = green leaves
- Let **g** = purple leaves

There are several things to note from the above representation of the tomato dihybrid cross:

1. Each gamete has one allele for fruit color and one for leaf color.

2. There are four different types of gametes possible for each F_1 heterozygote – each fruit color allele can be found in a gamete with either of the two leaf color alleles.

3. There are four different gametes possible for each F_1 male and also for each F_1 female. This means that there are 16 possible fertilization outcomes to produce the F_2 generation.

4. These 16 possible fertilizations result in nine different genotypes determining the four different F_2 phenotypes (Table 4.1).

5. These four different F_2 phenotypes occur in a ratio of 9 : 3 : 3 : 1, with the different classes distributed as shown in Table 4.2 with regard to dominance and recessiveness.

6. If we consider each trait separately (i.e. just fruit or leaf color), then it can be seen that each trait is inherited in the F_2 generation in the 3 : 1 ratio predicted by Gregor Mendel's **Law of Segregation**. In the Punnett square shown in Figure 4.2 there are 12 red fruiters to four yellow fruiters and, likewise, 12 green-leaved plants to four purple-leaved plants. This shows that

TABLE 4.1 A summary of F_2 genotypes and corresponding phenotypes

GENOTYPE	PHENOTYPE
RRGG RrGG RRGg RrGg	Red fruiting, green leaved
RRgg Rrgg	Red fruiting, purple leaved
rrGG rrGg	Yellow fruiting, green leaved
rrgg	Yellow fruiting, purple leaved

the inheritance of the alleles of one gene is unaffected by the inheritance of alleles at the other locus. Mendel realized this independence and encapsulated it in his second law: **The Law of Independent Assortment**. Using modern genetic terminology this can be stated as:

Each pair of alleles segregates independently so that in the gametes one member of each pair is equally likely to appear with either of the two alleles of the other pair.

TABLE 4.2 Dominant/recessive phenotype distribution among the F_2 generation in a dihybrid cross (traits 1 and 2 refer to arbitrary traits)

FERTILIZATION OUTCOME OF F_1 CROSS	PHENOTYPIC EXPRESSION OF TRAIT 1	PHENOTYPIC EXPRESSION OF TRAIT 2
9/16	Dominant	Dominant
3/16	Dominant	Recessive
3/16	Recessive	Dominant
1/16	Recessive	Recessive

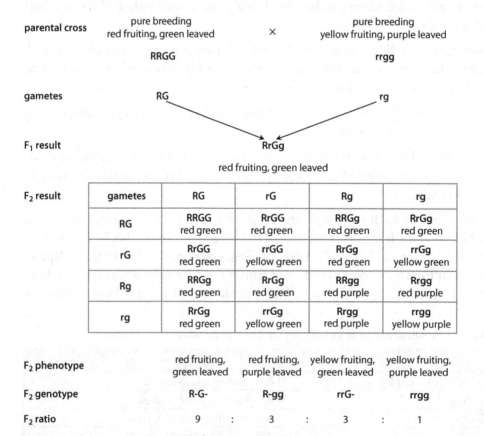

Figure 4.2 Genotypes and phenotypes of the tomato dihybrid cross.

4.2 Test cross

Our tomato grower was delighted to know how to guarantee production of his yellow-fruiting, purple-leaved plants. Any such plant was homozygous recessive and therefore pure breeding. He also wanted to ensure a supply of red-fruiting, green-leaved plants for commercial tomato production. To this end, he had discovered that he also needed pure-breeding plants. In the F_2 generation above, the red-fruiting, green-leaved plants can be one of four different genotypes (**RRGG**, **RrGG**, **RRGg**, and **RrGg**). To distinguish between these, and so identify the pure-breeding plant (**RRGG**), the tomato grower needs to do some test crosses. You may remember from Chapter 2 (Section 2.9) that this involves crossing plants of dominant phenotype, but unknown genotype, with homozygous recessive individuals. The phenotypes among the progeny indicate the genotypes of the phenotypically dominant plants. In this example, each F_2 red-fruiting, green-leaved plant needs to be crossed with a double homozygous recessive plant (**rrgg**), which will be yellow fruiting and purple leaved. The possible outcomes, depending upon the genotypes of the red-fruiting, green-leaved plant, are shown in Figure 4.3. Thus, to ensure a guaranteed supply of red-fruiting, green-leaved plants the tomato grower needs to keep seeds from plants yielding 100% red-fruiting, green-leaved plants when crossed with the double recessive.

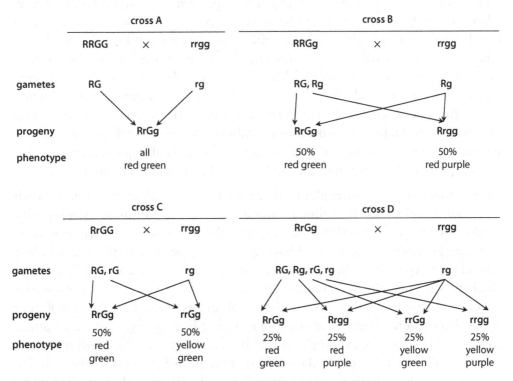

Figure 4.3 Possible outcomes of test crossing an F_2 red-fruiting, green-leaved plant.

This demonstration of the outcomes of various test crosses has also identified another useful Mendelian ratio. If a cross of a phenotypically dominant individual with a recessive homozygote yields progeny equally distributed among four phenotypic classes in a 1 : 1 : 1 : 1 ratio, this indicates that the individual of unknown genotype is a double heterozygote (here, **RrGg**).

4.3   Two genes – one trait

The previous two sections have established some important principles relating to the simultaneous inheritance of two genes, where each gene determines a different trait:

Gene A → trait 1

Gene B → trait 2

It is not, as stated earlier, uncommon to find a single characteristic under the influence of two genes:

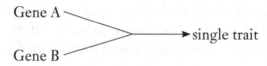

The principles established from looking at patterns of dihybrid inheritance, when two traits are involved, can be usefully applied to the second situation to enable various relationships between alleles at two different loci to be deduced. Clues come from looking at the ratios of phenotypic types among the progeny in the F_2 generation of a standard Mendelian cross. Two independently assorting genes typically produce a 9 : 3 : 3 : 1 ratio among four different phenotypes. Interactions between the alleles at two loci can result in modifications to this standard ratio. Different modifications indicate different kinds of interactions between the genes. These will be discussed in this and subsequent sections. This section deals with an example where the 9 : 3 : 3 : 1 F_2 ratio is maintained.

One of the first pieces of evidence that a trait can be influenced by more than one gene was obtained by William Bateson and Reginald Punnett working with poultry during the early 1900s, shortly after the rediscovery of Mendel's work. Different breeds of domestic fowl have different shaped combs (Figure 4.4). In a now classic experiment, Bateson and Punnett crossed rose-combed Wyandotte chickens with pea-combed Brahma chickens. All their F_1 chickens possessed a novel comb shape – walnut. When these walnuts were interbred, four different comb shapes were observed among the resulting F_2 progeny: rose, pea, walnut, and the distinctive spiked single comb already known in the Leghorn breed (Figure 4.5). The numerical relationships among the F_2 offspring immediately gave clues to the genetic basis of comb shape. The 9 : 3 : 3 : 1 ratio of walnut-,

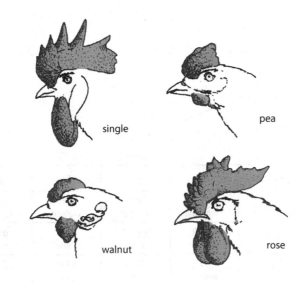

Figure 4.4 Fowl comb shapes.

parental cross	rose comb	×	pea comb

F₁ result		all walnut combed	

F₂ result	9 walnut :	3 rose :	3 pea :	1 single

Figure 4.5 Bateson and Punnett's cross between rose- and pea-combed fowl.

pea-, rose-, and single-combed fowl indicated that two independently assorting genes controlled expression of comb shape.

The two genes controlling comb shape are designated **R** and **P**. Table 4.3 shows the relationship between the different genotypes and the four different comb shapes, and Figure 4.6 represents Bateson and Punnett's original experiment.

TABLE 4.3 Relationships between genotypes and comb shape (**P/p** and **R/r** = alleles at the two relevant loci)

GENOTYPE	COMB SHAPE
P-R-	Walnut: at least one dominant allele at each locus
pprr	Single: double homozygous recessive
P-rr	Pea: at least one dominant **P** allele
ppR-	Rose: at least one dominant **R** allele

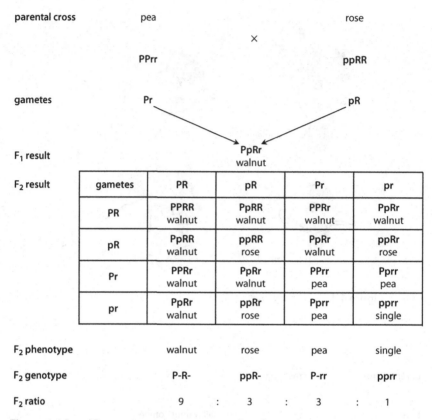

Figure 4.6 F_1 and F_2 genotypes and phenotypes resulting from crossing pure-breeding pea-combed fowls with pure-breeding rose-combed fowl. The F_2 generation is produced by interbreeding F_1 male and female walnut-combed fowl.

This section has described a situation where the products of two independently assorting genes interact to influence expression of a single trait. Each genotype class (**P-R-**, **P-rr**, **ppR-**, and **pprr**) determines a different phenotype. This is not always the case. Sometimes three, or even just two, phenotypes are observed in the F_2 generation in ratios such as 9 : 3 : 4 or 15 : 1. We know, however, that we are still dealing with two independently assorting genes as the F_2 ratio is clearly a modification of the standard 9 : 3 : 3 : 1. The observed phenotypes are in proportions of 16. From the way the 9 : 3 : 3 : 1 ratio is modified it becomes possible to deduce something about the nature of the interaction of the genes.

4.4 Complementary genes

The inheritance of comb shapes in different breeds of chicken was not Bateson and Punnett's sole interest. They were also intrigued by the genetic control of flower color in the sweet pea, *Lathyrus odoratus*. Normally the sweet pea produces purple flowers, but Bateson and Punnett had two different pure-breeding white varieties. Unexpectedly, when they crossed plants from these two white lines,

parental cross	pure breeding red flowering plants	×	pure breeding white flowering plants
F₁ result		all purple flowering plants	
F₂ result	purple flowering plants		white flowering plants
F₂ ratio	9	:	7

Figure 4.7 The results of Bateson and Punnett's cross between two white-flowering sweet pea varieties. The F₁ purple-flowering plants were selfed.

all the F_1 progeny were purple. Furthermore, self-pollination of these F_1 plants produced an F_2 generation of both purple- and white-flowering plants in a ratio of 9 : 7 purple- to white-flowering plants (Figure 4.7).

The 9 : 7 F_2 ratio provides the clue to the genetic basis of sweet pea flower color. It indicates that two genes are involved in determining petal color. The 9 : 7 ratio is a modification of the basic 9 : 3 : 3 : 1 ratio of a dihybrid cross. The genotypes that would result in the three separate phenotypes encompassing the "3 : 3 : 1" of the 9 : 3 : 3 : 1 ratio produce, instead, just one phenotypic class. In other words, to produce color in sweet pea flowers, a dominant allele must be present at each locus; otherwise, the petals are white. This can be seen in Figure 4.8.

parental cross	white flowering		×		white flowering	
	CCpp				ccPP	

F₁ result			purple flowering CcPp			

F₂ result	**gametes**	**CP**	**cP**	**Cp**	**cp**
	CP	CCPP purple	CcPP purple	CCPp purple	CcPp purple
	cP	CcPP purple	ccPP white	CcPp purple	ccPp white
	Cp	CCPp purple	CcPp purple	CCpp white	Ccpp white
	cp	CcPp purple	ccPp white	Ccpp white	ccpp white

F₂ phenotypic ratio	9 purple-flowering plants : 7 white-flowering plants

Figure 4.8 F₁ and F₂ genotypes and phenotypes resulting from crossing two pure-breeding, white-flowering sweet pea varieties. The F₂ generation is produced by interbreeding the F₁ purple-flowering plants. (The two loci determining flower color are designated **C** and **P**.)

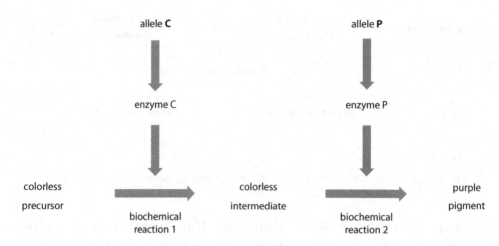

Figure 4.9 Producing purple sweet pea pigment requires two biochemical reactions, each catalyzed by a different enzyme.

We can produce a biochemical explanation for the determination of sweet pea petal color. Two consecutive biochemical reactions change a colorless precursor into a colored pigment. Each reaction requires a different enzyme, the first encoded by gene **C** and the second by gene **P** (Figure 4.9). Only those plants with active forms of both enzymes, individuals with genotypes **C-P-**, will be purple. All other genotypes lack at least one enzyme and so result in white flowers (Table 4.4).

The two genes determining sweet pea flower color are said to be complementary. This term reflects the fact that functional products of each locus are needed for full phenotypic expression. This only occurs when both loci are present in the dominant state, resulting in products that work in a sequential fashion to complete the necessary biochemical pathway to produce a particular phenotype.

TABLE 4.4 The relationship between genotype and sweet pea flower color

GENOTYPE	ACTIVE ENZYME PRESENT	SWEET PEA FLOWER COLOR
C-P-	Enzymes C and P	Purple
C-pp	Enzyme C	White
ccP-	Enzyme P	White
ccpp	None	White

Complementary gene actions can be recognized when the $9 : 3 : 3 : 1$ F_2 phenotypic ratio is modified to $9 : 7$.

Complementary gene products work sequentially. Occasionally, a situation is identified where gene products work in parallel; that is, there are two ways of achieving the same result! Examples of this are given at the end of Section 4.5.

4.5   Epistasis

Another kind of gene interaction that results in modified $9 : 3 : 3 : 1$ F_2 ratios occurs when alleles at one locus suppress the expression of alleles at a second locus. This effect is observed between the alleles of two genes determining the coat color of Labrador and retriever dogs. A dominant allele (**B**) at the first locus results in black coats; the recessive homozygote (**bb**) is brown. However, another gene can stop production of either black or brown fur. There is no effect when the dominant allele (**E**) is present at the second locus, but dogs homozygous recessive at this locus (**ee**) will be golden-brown. This phenomenon, where one locus influences expression of a second, is referred to as **epistasis**. The coat color example illustrates **recessive epistasis** because it is when recessive alleles are present at the second locus that expression of fur color is suppressed. In other situations it can be the dominant allele at an epistatic locus modifying expression at a second locus.

The tell-tale ratio indicating recessive epistasis is a $9 : 3 : 4$ ratio in the F_2 generation. Consider a mating between a pure-breeding black and a pure-breeding golden Labrador dog (Figure 4.10). The phenotype of the F_1 generation could suggest a monohybrid situation, but the three phenotypic classes of the F_2 (black, brown, and golden fur color), present as proportions of 16, unambiguously confirm two genes are involved. The $9 : 3 : 4$ ratio indicates recessive epistasis with the "4" encompassing the "3 : 1" classes of the $9 : 3 : 3 : 1$ segregation ratio.

Epistatic suppression of loci determining coat color is a common phenomenon. For example, albinism regularly occurs in many mammal species (including humans), and has also been reported among birds, fish, and snakes. Pale skin or white coat is generally the result of a recessive epistatic allele, when the recessive homozygote prevents the expression of pigment encoding alleles at another locus. Occasionally, however, a dominant epistatic allele prevents the expression of color at a separate locus. Examples of such dominant epistasis are not confined to animal coloration.

Squash fruits can be white, yellow, or green. A dominant allele at one locus results in yellow fruit, while the recessive homozygote is green. These fruit will be yellow or green as long as another locus is homozygous recessive. A dominant allele at this second locus abolishes all color in the squash fruit (Table 4.5).

parental cross	pure breeding black furred	×		pure breeding golden furred
	BBEE			**bbee**

F₁ result		black furred BbEe		

F₂ result

gametes	BE	Be	bE	be
BE	BBEE black	BBEe black	BbEE black	BbEe black
Be	BBEe black	Bbee yellow	BbEe black	Bbee yellow
bE	BbEE black	BbEe black	bbEE brown	bbEe brown
be	BbEe black	Bbee yellow	bbEe brown	bbee yellow

F₂ phenotype	black furred	brown furred	golden furred
F₂ genotype	B-E-	bbE-	B-ee and bbee
F₂ ratio	9 :	3 :	4

Figure 4.10 Investigating the inheritance of Labrador fur color. A pure-breeding black male was crossed with a pure-breeding golden female. A F₁ black female was mated with a F₁ black male from the same litter. (**B/b** = black/brown fur alleles and **E/e** = alleles at the epistatic locus.)

Figure 4.11 shows the consequences of crossing pure-breeding white- and green-fruiting squashes. A 12 : 3 : 1 F₂ ratio is a common clue to the operation of dominant epistasis.

Epistasis does not always involve two genes acting in opposition to each other. Sometimes two different genes can each have the same role in determining a trait and can substitute for each other. For example, generally the shape of the seed capsule of the common weed shepherd's purse, *Capsella bursa-pastoris*, is triangular. Occasionally, ovoid capsules are produced, but only when recessive alleles

TABLE 4.5 Relationship between genotype and squash fruit color (**C/c** = alleles of the color-producing gene and **S/s** = alleles of the suppressing locus)

GENOTYPE	PHENOTYPE
C-ss	Yellow
ccss	Green
C-S-	White
ccS-	White

parental cross	pure breeding white fruiting		×		pure breeding green fruiting
	CCpp				ccPP

F₁ result — all white fruiting CcSs

gametes	CS	cS	Cs	cs
CS	CCSS white	CcSS white	CCSs white	CcSs white
cS	CcSS white	ccSS white	CcSs white	ccSs white
Cs	CCSs white	CcSs yellow	CCss yellow	Ccss yellow
cs	CcSs white	ccSs white	Ccss yellow	ccss green

F₂ phenotype: white fruiting   yellow fruiting   green fruiting

F₂ genotype: C-S- and ccS-   C-ss   ccss

F₂ ratio: 12 : 3 : 1

Figure 4.11 The inheritance of squash fruit color. Pure-breeding white-fruiting and pure-breeding green-fruiting plants were crossed. The resulting F₁ white-fruiting plants were selfed. (**C/c** = alleles of the color-producing gene and **S/s** = alleles of the suppressing locus.)

are present at two specific loci, referred to as T_1 and T_2. These two genes control key stages in two different developmental pathways, either of which can result in triangular seeds. Thus, it is only when neither pathway is operational (i.e. a plant is a double homozygote) that ovoid seeds are produced (Table 4.6). The evidence for this duplicate gene situation came from crosses between double heterozygous plants when progeny are produced in a ratio of 15 : 1 triangular- to ovoid capsule-producing plant. An ovoid phenotype is only produced by the double homozygous recessive (Figure 4.12).

TABLE 4.6 The determination of seed capsule shape in shepherd's purse by duplicate genes T_1 and T_2: relationship between genotype and capsule shape

GENOTYPE	GENE PRODUCT	PHENOTYPE
$T_1T_1T_2T_2$	Active T_1 and T_2	Triangular
T_1-t_2t_2	Active T_1/inactive T_2	Triangular
$t_1t_1T_2$-	Inactive T_1/active T_2	Triangular
$t_1t_1t_2t_2$	Inactive T_1 and T_2	Ovoid

| double heterozygotes | triangular $T_1t_1T_2t_2$ | | × | triangular $T_1t_1T_2t_2$ | |

F$_2$ result	gametes	T_1T_2	T_1t_2	t_1T_2	t_1t_2
	T_1T_2	$T_1T_1T_2T_2$ triangular	$T_1T_1T_2t_2$ triangular	$T_1t_1T_2T_2$ triangular	$T_1t_1T_2t_2$ triangular
	T_1t_2	$T_1T_1T_2t_2$ triangular	$T_1T_1t_2t_2$ triangular	$T_1t_1T_2t_2$ triangular	$T_1t_1t_2t_2$ triangular
	t_1T_2	$T_1t_1T_2T_2$ triangular	$T_1t_1T_2t_2$ triangular	$t_1t_1T_2T_2$ triangular	$t_1t_1T_2t_2$ triangular
	t_1t_2	$T_1t_1T_2t_2$ triangular	$T_1t_1t_2t_2$ triangular	$t_1t_1T_2t_2$ triangular	$t_1t_1t_2t_2$ ovoid

F$_2$ phenotype   triangular   ovoid

F$_2$ ratio   15   :   1

Figure 4.12 The inheritance of shepherd's purse capsule shape. Pure-breeding plants for triangular capsules of two different genotypes (**$T_1T_1t_2t_2$** and **$t_1t_1T_2T_2$**) were crossed. The resulting F$_1$ plants, with triangular capsules, were selfed.

The shepherd's purse capsule shape provides an example of one phenotype produced by the expression of either of two alternative genes. The reverse situation sometimes exists – when one gene influences several phenotypes (**pleiotropy**). Consider the gene causing phenylketonuria (PKU) in humans. The main phenotype is mental impairment. The defect in the detoxification of the amino acid phenylalanine also interferes with the synthesis of the skin pigment melanin. Hence, PKU sufferers also often share the phenotype of pale skin and blond hair.

4.6   The chi-squared (χ^2) test

This chapter has considered a range of ways that the gene products from two loci interact to determine a trait. Clues to the type of interaction come from analyzing the frequencies, or ratios, of different phenotypic expressions among the progeny of crosses, summarized in Table 4.7. If, for example, three different phenotypes, in a ratio of $9:3:4$, are observed in the F$_2$ generation, we conclude that two genes are involved, with recessive epistasis occurring between them, while progeny showing a frequency relationship of $12:3:1$ suggests dominant epistasis.

To be able to draw conclusions about the nature of the genes' interactions, we have to be confident that our data do indeed resemble one of these useful dihybrid ratios, and that any observed deviations from the expected ratios are the result of the random fluctuations that are the inevitable consequences of the

TABLE 4.7 Gene interactions and modified Mendelian F_2 ratios

TYPE OF GENE INTERACTION	F_2 GENOTYPIC CLASSES AND RESULTING PHENOTYPIC RATIOS			
	A-B-	aaB-	A-bb	aabb
None	9	3	3	1
Complementary genes	9		7	
Recessive epistasis (**aa** suppressing expression of **B/b**)	9	3	4	
Dominant epistasis (**B** suppressing expression of **A/a**)	12		3	1
Duplicating genes	15			1

chance nature of fertilization and the segregation and independent assortment of alleles into gametes. We can employ a simplest statistical test, the chi-squared (χ^2) test, to assess whether any observed deviations from the expected progeny ratios can be attributed solely to chance or are caused by something else. As with all statistical tests, the following procedure is used:

1. A null hypothesis (H_0) is first formulated. This states that there is no significant difference between the measured values (here, the observed frequencies of different phenotypes) and the predicted ones (e.g. based on a phenotypic frequency ratio of $9 : 3 : 3 : 1$).

2. A value is then calculated for the test statistic, in this case for χ^2 (Box 4.1).

3. Once a calculated χ^2 value has been obtained it is possible to make a decision to accept or reject the null hypothesis, and thus to know whether any

---

BOX 4.1 CALCULATING A VALUE FOR χ^2

The test statistic χ^2 directly compares observed (O) and expected (E) values:

$$\chi^2 = \sum \frac{(O-E)^2}{E}$$

For each frequency component (e.g. each phenotypic class):

(i) The expected frequency value (E) is subtracted from the observed frequency value (O).

(ii) The ($O - E$) value is squared and divided by the expected frequency value (E).

(iii) The values observed for each component are summed ($\sum$).

This produces a **calculated value** for χ^2.

BOX 4.2 ACCEPTING OR REJECTING THE NULL HYPOTHESIS

Once a calculated value for χ^2 has been obtained, a **significance level** and the **degrees of freedom** are fixed.

The **significance level** (***P***) represents the probability that the null hypothesis is correct. For analysis of genetic data, a significance level of 0.05 is commonly selected. This means for 95% of the results, agreement would be expected between the observed and predicted frequency values, but it is acceptable for there to be deviations in 5% of cases.

The **degrees of freedom (d.f.)** take into account the number of different data categories (here, phenotypic classes), because more deviation from the expected result occurs with a greater number of categories. The number of degrees of freedom is equal to the number of data categories minus one.

Once the significance level and the number of degrees of freedom are fixed, a comparison is made between the calculated and **critical value** for χ^2. **Probability tables** exist of critical values for χ^2 (see Appendix) at different significance levels and degrees of freedom. The null hypothesis is accepted if the calculated value is less than the critical value and is rejected if the calculated value is greater than the critical value.

difference between an observed and theoretical ratio is due to chance, or has some other cause (Box 4.2).

An example is given in Box 4.3 to show how the χ^2 test is used.

Summary

- Dihybrid crosses investigate the patterns of inheritance when two genes are involved.

- The segregation and subsequent inheritance of alleles at each gene occur independently.

- When two genes control variation in two different characteristics, a phenotypic ratio of 9 : 3 : 3 : 1 is obtained in the F_2 generation.

- The 9 : 3 : 3 : 1 F_2 ratio may be modified when both genes control the same characteristic. The nature of the modified ratio indicates the type of interaction between the two loci.

- Complementation occurs when a functional gene product of both loci is required for full expression of a character.

- Epistasis refers to the situation when either the dominant or the recessive allele at one locus prevents expression at the second.

- A χ^2 statistical test assesses whether observed data match predicted results.

BOX 4.3 CALCULATING A χ^2 VALUE USING THE DATA FROM A GENETIC CROSS

Section 4.2 described the use of a test cross to identify, for the tomato grower, a supply of seed that was guaranteed to produce red-fruiting, green-leaved plants. In one test cross, the following results were obtained:

	Number of plants
Red fruits, green leaves	135
Red fruits, purple leaves	163
Yellow fruits, green leaves	155
Yellow, fruits purple leaves	148

These results suggest a 1 : 1 : 1 : 1 ratio, which would indicate that the red-fruiting, green-leaved plant was heterozygous, not the double homozygote the tomato grower had wanted. To confirm this conclusion, a χ^2 test can be performed on the results:

Phenotype	Observed numbers (O)	Expected numbers (E)	$O - E$	$(O - E)^2$	$(O - E)^2/E$
Red fruits, green leaves	135	150	−15	225	1.5
Red fruits, purple leaves	163	150	13	169	1.13
Yellow fruits, green leaves	155	150	5	25	0.17
Yellow fruits, purple leaves	148	150	−2	4	0.03
Total	600	600			2.83

- Null hypothesis: there is no difference between observed results and an expected ratio of 1 : 1 : 1 : 1

- Significance level = 0.05; degrees of freedom = 3; calculated χ^2 value = 2.83; critical χ^2 value = 7.82 (see Appendix for table of χ^2 statistics)

As the calculated χ^2 value is less than the critical χ^2 value, at a significance level of 0.05 and three degrees of freedom, the null hypothesis can be accepted.

The results, therefore, resemble a 1 : 1 : 1 : 1 ratio and we can conclude that the red-fruiting, green-leaved plant was heterozygous.

Problems

1. In the tomato, purple stems are dominant to green and the presence of hairs on the stems is dominant to their absence. A purple, hairy-stemmed plant was self-fertilized and produced the following offspring: 133 purple hairy, 47 purple hairless, 45 green hairy, and 15 green hairless. What was the genotype of the self-fertilized purple hairy-stemmed plant?

2. Fur color in mice may be black (**B**) or brown (**b**) and their tails may be short (**S**) or long (**s**). What proportion of the progeny of a cross between two heterozygotes (**BbSs**) will have brown hair and short tails, if fur color and tail length assort independently?

3. Two pure-breeding blind strains of crickets were crossed to each other. The F_1 crickets all showed normal sight. When these F_1 crickets were crossed among each other, the resulting F_2 progeny consisted of 178 crickets with normal sight and 142 blind crickets.

   (a) Explain the results.

   (b) What fraction of the F_2 generation would you expect to be pure breeding for sightedness?

   (c) What phenotypes would you expect among the progeny, and in what proportions, if the F_1 crickets were crossed to a double homozygous recessive individual?

4. In Shorthorn cattle, polled (hornless) is dominant to horned. Coat color may be red, white, or a mixture of red and white hairs referred to as roan; the two colors being determined by two separate alleles at a single locus. A roan, horned bull was mated on four different occasions with a white, polled cow. All the resulting calves were polled, but five were white and four were roan. What were the genotypes of the parents?

5. Removal of infected honey bee larvae from a hive is under genetic control. It involves (i) uncapping the cell of an infected larvae and (ii) removing the infected larvae. Successful uncapping and removal depends upon appropriate alleles being present at two loci:

Genotype	Uncapping	Removal
U-R-	No	No
uuR-	Yes	No
U-rr	No	No
uuR-	Yes	Yes

   One hundred worker bees were taken at random from a hive, and their uncapping and removal behavior investigated. Only six bees successfully uncapped and removed infected larvae. A further 20 uncapped, but did not remove larvae, while the remaining 74 exhibited no nest-cleaning behavior. Suggest genotypes for these 100 bees.

6. Two varieties of corn produce colorless aleurone layers in their seeds. When crossed, the resulting seeds all possessed a purple aleurone layer. The same color also appeared in 270 of the 480 F_2 plants. How do you explain inheritance of a purple aleurone layer in corn?

7. Cyanogenesis is an anti-predation mechanism in the white clover, *Trifolium repens*. It is regulated by two loci. The first (**Ac/ac**) ensures production of glucoside and the second (**Li/li**) encodes the enzyme linamarase (β-D-glucosidase), which catalyzes the production of hydrogen cyanide from the glucoside.

Ac/ac   Li/li

Precursor → glucoside → hydrogen cyanide

Thus, if the two loci are in the dominant state (**Ac-Li-**), leaves readily produce hydrogen cyanide when crushed. If only the **Ac** locus is producing a functional product, cyanogenesis still occurs, but at a slow, spontaneous rate. The plant is acyanogenic if the first gene product is absent. From a cross involving plants heterozygous at both loci, what is the probability that any one of the resulting clover plants will be (a) cyanogenic; (b) acyanogenic; (c) show slow cyanogenesis?

8. In horses, a trotting gait (**T**) is dominant to a pacing gait (**t**) and black coat (**B**) is dominant to chestnut (**b**). What are the genotypes and phenotypes of the parents which produce offspring in the following proportions: 1/8 chestnut pacers, 1/8 chestnut trotters, 3/8 black trotters, and 3/8 black pacers?

9. Achondroplasia (an inherited form of dwarfism) and short sightedness are both dominantly inherited human traits. A man with achondroplasia marries a woman who is short sighted. If both individuals are heterozygous:

   (a) What is the probability that any child has achondroplasia or is short sighted?

   (b) What is the probability that a child is both short sighted and has achondroplasia?

10. A cat breeder crossed two white cats expecting all the kittens to be white. She was greatly surprised to obtain two tabby and a black kitten among the litter. The next time the white female came into season, the breeder repeated the mating. Again, the litter contained black and tabby kittens. A friend suggested that the two white cats might each possess a dominant color-suppressing gene and that, when the cats were mated, non-suppressing alleles had segregated to some of the kittens, enabling different colors to be expressed. To test this hypothesis the breeder did one final mating. Collectively, the three litters had yielded 17 white, five tabby, and two black kittens. Use a χ^2 test to decide whether the breeder's friend was correct.

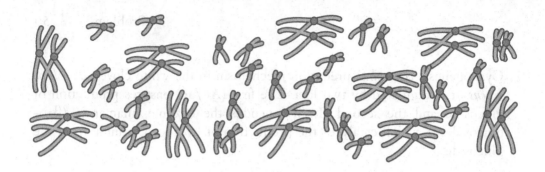

The Chromosomal Basis of Inheritance

CHAPTER 5

The rediscovery of Mendel's work on the nature of inheritance in 1900 led to a great burst of genetic experimentation. It quickly became clear that Mendel's laws were valid for a wide variety of organisms. Geneticists also realized that the inheritance patterns of Mendel's factors paralleled the behavior of the newly discovered chromosomes during sexual reproduction. As a consequence, the **chromosome theory of inheritance** was proposed – that chromosomes were the carriers of Mendel's factors, or "genes," as they soon became known.

This chapter considers

- The physical nature of chromosomes

- The transmission of chromosomes from cell to cell and from generation to generation during the processes of mitosis and meiosis, respectively

- The correlation between the segregation patterns of genes, as discussed in Chapters 2–4, and the behavior of chromosomes during meiosis

5.1   The structure of chromosomes

A chromosome is one long molecule of DNA (see Chapter 11) arranged within a framework of protein molecules. Chromosomes can only be seen as discrete structures during the two processes of nuclear division – **mitosis** and **meiosis** – when they become highly condensed. It is, therefore, only during mitosis and meiosis that it is possible to distinguish the size, shape, and number of chromosomes within a cell. Figure 5.1 shows a scanning electron micrograph of human chromosomes during mitosis, when each chromosome is a double structure, composed of a pair of **chromatids**, held together at the **centromere**.

The frequently used terms "chromosome" and "chromatid" need to be clarified. Within a normal working eukaryotic cell each chromosome is present as a single structure. When a cell divides it is essential that each new cell has a complete and accurate copy of an organism's genetic information. Thus, prior to division, each chromosome makes an exact copy of itself. During the early stages of both mitosis and meiosis the two new chromosomes remain attached at a site called the centromere, visible under the microscope as a constriction (Figure 5.1). While the two replicas are attached they are known as chromatids. Once in separate nuclei they are again referred to as chromosomes. When a cell is not dividing, the chromosomes exist in a partially unwound state referred to as **chromatin** (Section 11.4).

Often a **karyotype**, a display of a cell's chromosomes ordered according to size, is made (Figure 5.2). This enables a detailed analysis of a cell's chromosomes. A digital image is taken of a cell arrested in the middle of mitosis. The technique involves rupturing the nucleus, which results in a "spread" of chromosomes that then are imaged. A karyotype is produced from the image. Chromosomes are either cut out or digitally moved, and then arranged in pairs according to size,

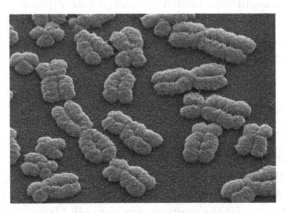

Figure 5.1 Scanning electron micrograph of human chromosomes. Courtesy of Science Photo Library.

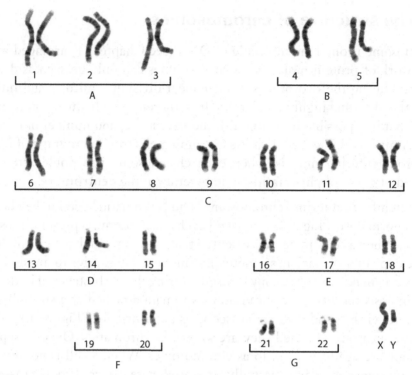

Figure 5.2 Karyotype of a G-banded human male. Autosomes are places in six groups, A–G, according to size.

banding patterns and centromere position. Most species show variation in the sizes of the different chromosomes. A group of similar-sized chromosomes (e.g. the chromosomes of group D in the human karyotype; Figure 5.2), can be sorted into matching pairs because each different chromosome possesses a distinctive banding pattern when stained with chemical dyes (Box 5.1). Two chromosomes – the **sex chromosomes**, which determine the sex of an individual – are usually displayed separately at the end of a karyotype. This enables an individual's gender to be quickly recognized. Human females, for example, possess two identical **X chromosomes**, while males have one X and a much smaller **Y chromosome** (for more details about sex chromosomes, see Chapter 6). All other chromosomes are known as **autosomes**.

5.2   The number of chromosomes

The karyotype in Figure 5.2 shows that each human cell has 46 chromosomes. The number of chromosomes is constant for all **somatic** cells of a given species, although it can vary enormously between species. For example, each cell of the scorpion possesses just four chromosomes, while some ferns, such as the adder's tongue, have more than 1000 chromosomes per cell! More common, however, is a chromosome number between 10 and 50 (Table 5.1).

Until the late 1960s, mitotic chromosomes could only be distinguished under the light microscope on the basis of their relative size and the positions of their centromeres. Thus, the situation often arose that two or more chromosomes could not be identified separately. During the late 1960s, several different research groups developed techniques that produced differential staining along the long axis of chromosomes. The results were recognizable and reproducible banding patterns for the different chromosomes of a species. Several of the techniques used **Giemsa stain**. Depending upon the conditions associated with the use of the stain, different distinctive patterns are obtained, referred to as C, **G**, and **R** banding. The variety of staining reactions, under different conditions, reflects the heterogeneity and complexity of chromosome composition. With G banding, **heterochromatin** (genetically inactive chromatin) is dark staining and **euchromatin** (transcriptionally active DNA) is lighter staining. R (reverse banding) shows the reverse situation. C banding highlights centromeres. The chromosomes' bands obtained by these various treatments enable identification of normal and abnormal chromosomes. They also are informative about aspects of chromatin structure and the organization of the genome.

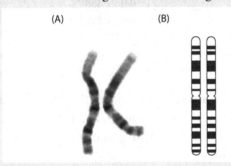

(A)   (B)

Box 5.1 Figure 1 Human chromosome 1 (represented as a pair of chromatids) showing the main G bands produced with Giemsa stain. Left: a photomicrograph; right: an ideogram (a black and white representation) of chromosome 1.

5.3 Homologous pairs of chromosomes

The chromosomes of a somatic cell can be arranged in matching pairs (Figure 5.2). These pairs are referred to as **homologous pairs**. The 46 chromosomes of a human somatic cell, therefore, represent two **sets** of chromosomes. One of these sets originally came from the mother (the **maternal set**) and the other from the father (the **paternal set**). Each set contains 23 different chromosomes: 22 **autosomes** (non-sex-determining chromosomes) and a sex chromosome. The two chromosomes of a homologous pair have the same size, shape, and functions. They have a characteristic set of genes that determine a collection of different

TABLE 5.1 Chromosome number in a selection of plant and animal species

ANIMAL		PLANT	
SPECIES	CHROMOSOME NUMBER	SPECIES	CHROMOSOME NUMBER
Homo sapiens (human)	46	*Pisum sativum* (garden pea)	14
Pan troglodytes (chimpanzee)	48	*Solanum tuberosum* (potato)	48
Equus caballus (horse)	64	*Triticum aestivum* (wheat)	42
Felis domesticus (cat)	38	*Zea mays* (maize)	20
Oryctolagus cuniculus (rabbit)	44	*Vicia faba* (broad bean)	12
Xenopus laevis (toad)	36	*Gossypium hirsutum* (cotton)	56
Musca domesticus (housefly)	12	*Sequoiadendron giganteum* (redwood)	22
Mus musculus (mouse)	40	*Solanum lycopersicum* (tomato)	24
Alligator mississippiensis (alligator)	32	*Antirrhinum majus* (snapdragon)	8
Carssinus auratus (goldfish)	94	*Oryza sativa* (rice)	24

traits. A particular gene will always be found at the same **locus**, or position, on a given chromosome and on its homologous partner. Much research is currently directed towards determining, or **mapping**, the precise locations of the nearly 22,000 human genes currently identified.

5.4   Diploid and haploid cells

Any cell containing two sets of chromosomes is described as **diploid** and sometimes represented as **2n**. Gametes generally have just one set of chromosomes. These are referred to as **haploid** cells and represented as **n**. The genetic information present in a haploid set of chromosomes is also referred to as an organism's **genome**. The modern discipline of **genomics** represents the study of a species' genome, usually from a molecular perspective. Gametes are not, however, the only haploid cells. Many plants show an alternation of two distinct generations within their life cycle. The cells of the spore-producing **sporophyte** generation are diploid, while those of the gamete-producing **gametophyte** generation are haploid (Figure 5.3). For most of their life cycle fungal cells are haploid.

It is important to remember that, although each pair of chromosomes in a diploid cell possesses the same sequence of genes, the two chromosomes are not usually

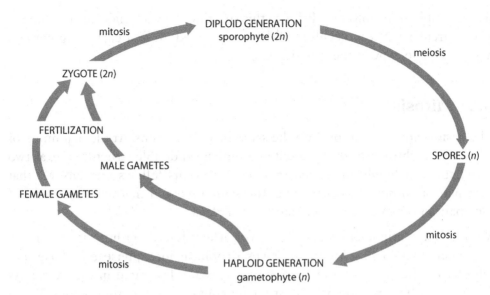

Figure 5.3 Alternation of haploid and diploid generations in plants.

identical. This is because individuals are heterozygous at many loci (i.e. possess a different allele on each chromosome). Those rare examples, where virtually all loci are homozygous, produce inbred populations, such as the African cheetah or the northern elephant seal.

5.5 Two types of nuclear division

As chromosomes carry hereditary information it is vital that, each time a cell divides during growth of an organism, both daughter cells receive a complete and uncorrupted copy of this vital information. A precisely controlled process of nuclear division called **mitosis** ensures that each daughter cell receives the same number and types of chromosomes as were present in the parental cell.

The life cycle of most eukaryotes includes sexual reproduction. There must, therefore, be a stage in the life cycle when the chromosome number is halved, so that gametes contain a haploid set of chromosomes. If that did not occur, then the chromosome number of a species would double each generation, at fertilization. **Meiosis** is the nuclear division resulting in cells with half the chromosome number of the parental cells. Meiosis is thus essential for maintaining the balanced diploid chromosome number of a species, as well as generating variation among offspring. The next four sections consider the processes of mitosis and meiosis in more detail.

Propagation of prokaryotic bacteria is achieved by a different mechanism – **binary fission**. Generally, when a bacterial cell has grown to twice its starting size, it simply splits into two. Prior to this splitting the bacteria's chromosome (a single circular DNA molecule) will have replicated, so that each daughter cell has a complete copy of the genetic material.

5.6 Mitosis

The basic features of mitosis are the same in all organisms. At the beginning of mitosis each chromosome is present as a replicated double structure (i.e. as two chromatids). A highly organized sequence of events follows that ensures that each pair of chromatids separates at the same time and that one chromatid of each pair finds its way into each daughter nucleus.

Mitosis is a continuous process. However, when describing mitosis, it is more convenient to recognize four main stages: **prophase**, **metaphase**, **anaphase**, and **telophase**. The key events of these four stages are illustrated in Figure 5.4. As discussed earlier (Section 5.2), most nuclei contain many chromosomes. When describing mitosis, the key features are more clearly conveyed by showing cells with a diploid chromosome number of four ($2n = 4$).

Prophase is the longest stage of mitosis. During this phase the long chromatin threads coil up, becoming shorter and thicker and recognizable as individual chromosomes. Other events are also occurring to make sure that there will be an equal and accurate distribution of chromosomes to the two daughter cells:

1. **Centrioles**, a collection of tiny microtubules, divide and move to opposite ends or **poles** of the cell.

2. Long fibers develop from the centrioles. Some of these **spindle fibers** extend from pole to pole, while others link up with the chromosomes and ensure a complete set of chromosomes segregates to each new nucleus. A star of short fibers, forming the **aster**, can also be seen radiating from each centriole.

Towards the end of prophase each chromosome is seen to consist of two chromatids: chromosome replication occurs before mitosis, but the two chromatids remain closely associated. Thus, early in prophase each chromosome is only visible as a single structure (Figure 5.4). As the individual chromatids become visible the nucleolus disappears and the nuclear envelope breaks up into a number of small vesicles.

The chromatid pairs now lie free in the cytoplasm. They gradually line up in the center of the cell, or **equator**. At this stage, **metaphase**, the pairs of chromatids are held at the equator by the attachment of spindle fibers to their centromeres. **Anaphase** soon follows. The centromeres divide and the attached spindle fibers contract, pulling the separated chromatids to opposite ends of the cell. The reverse of prophase now happens. The chromosomes uncoil to once again form

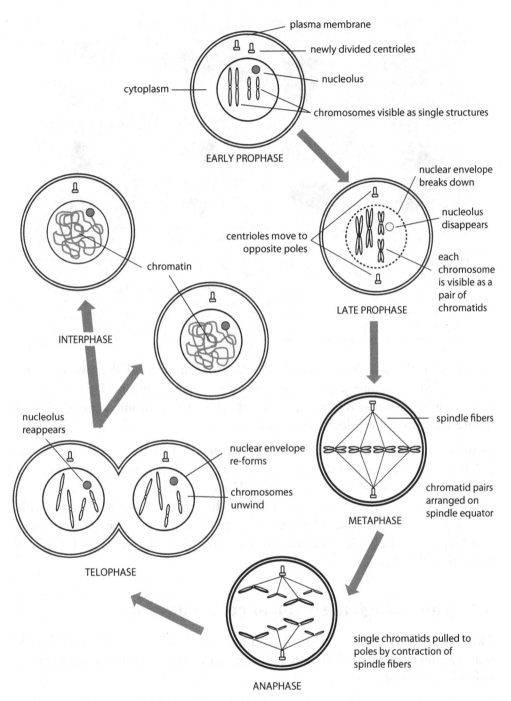

Figure 5.4 Mitosis in a generalized animal cell.

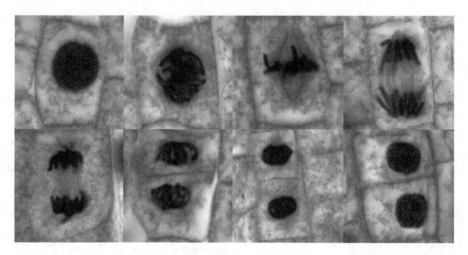

Figure 5.5 Light micrograph of root tip cells from an onion undergoing mitosis. Top row, from left to right: interphase chromosomes condense and appear as long thread-like structures (prophase). They then align along the center of the cell (metaphase). Each chromosome consists of two identical sister chromatids that separate and are pulled to opposite ends of the cell (anaphase). Bottom row, from left to right: nuclear membranes form around the two daughter nuclei as the chromosomes de-condense (telophase). The cell then divides (cytokinesis) and, as these are plant cell, a new cell wall is produced. Chromosomes extend and are indistinguishable as a mass of chromatin. Courtesy of Science Photo Library.

long diffuse threads. New nuclear envelopes form around the two separated groups of chromosomes among which a new nucleolus appears. This reorganization phase is known as **telophase**. The behavior of the chromosomes during the different stages can be clearly seen in Figure 5.5.

Mitosis is followed by cell division (**cytokinesis**). In animal cells this simply involves constriction of the cytoplasm between the two new nuclei. Dividing plant cells have to construct a new wall. Thus, vesicles containing wall material gather in the middle of the cell and fuse, producing the **cell plate** against which new walls are constructed. The period between two nuclear divisions is known as **interphase**. The cycle of events, from one mitosis to the next, is referred to as the **cell cycle** (Box 5.2).

5.7   The biological significance of mitosis

The two daughter nuclei produced by mitosis have the same number of chromosomes as the parental nucleus and are genetically identical. Mitosis occurs in the following situations:

- It enables the growth of an organism from a unicellular zygote to a multicellular individual.

- It generates nuclei, and so cells, for repair of damaged tissues and for normal cellular replacement during an organism's life (e.g. most vertebrates constantly replace cells in their skin and their gut lining).

BOX 5.2 THE CELL CYCLE

The life of a somatic cell is represented as a cycle, beginning and ending with mitosis (**M**). During interphase, a cell is performing its normal functions (e.g. as a bone or a leaf cell). G_1 and G_2 are the two "gap" phases. During G_1 the cell is growing and performing its normal metabolic functions, including preparation for cell division. At a certain point, referred to as the **restriction point**, the cell commits to division and moves into the **S phase**. During this "synthesis" phase the chromosomes are replicated in preparation for mitosis. All the proteins and other molecules needed for mitosis and cytokinesis are synthesized during the G_2 phase. During G_1, cells may enter G_0, when they cease to proliferate. This often indicates mature, terminally differentiated cells, such as nerve or heart muscles cells.

The lengths of the S and G_2 phases tend to be constant and species specific. For most of interphase, cells are in the G_1 phase, which can vary enormously in length – from hours in a growing embryo to years for a nerve cell. The G_1/S boundary is a major **checkpoint** in the cycle. Cytoplasmic signals are needed to enable a cell to proceed into an S phase and thus trigger events towards mitosis. Genes involved in checkpoints are often mutated in tumors.

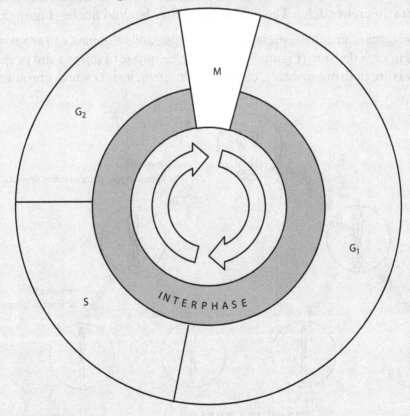

Box 5.2 Figure 1 The cell cycle.

Figure 5.6 An example of asexual reproduction: budding in hydra to produce a new organism.

- Some organisms reproduce asexually – a new individual is produced in a non-sexual process from just one parent (Figure 5.6). Mitosis is the basis of asexual reproduction, generating identical nuclei for new cells.

5.8 Meiosis

Meiosis produces nuclei with half the number of chromosomes of the parental nuclei. This is achieved by two divisions, referred to as meiosis I and meiosis II. **Meiosis I** is the "reduction division." It results in two daughter nuclei with half the number of chromosomes. These chromosomes, however, still consist of a pair of chromatids. During **meiosis II**, as in mitosis, the pair of chromatids are separated into different nuclei. The final result is four haploid nuclei (Figure 5.7).

The key stages of prophase, metaphase, anaphase, and telophase can again be recognized in each division (Figure 5.8). Initially, prophase of mitosis and prophase I of meiosis are indistinguishable, until the point in meiosis at which chromosomes

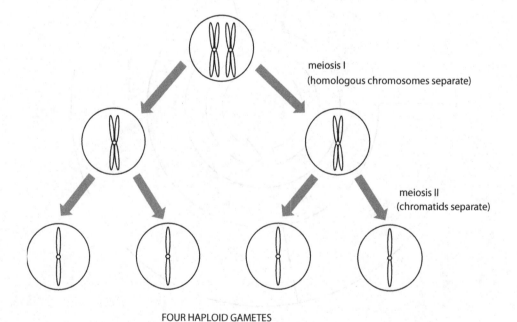

meiosis I
(homologous chromosomes separate)

meiosis II
(chromatids separate)

FOUR HAPLOID GAMETES

Figure 5.7 A summary of meiosis (one pair of homologous chromosomes is shown).

can be seen associated together in homologous pairs (Figure 5.8). This pairing of chromosomes during meiosis is known as **synapsis** and each homologous pair is called a **bivalent**. The close association enables sections of chromatids to be exchanged with corresponding sections of the homologous pair at points called **chiasmata**. These exchanges and their consequences are discussed in Section 7.3. Towards the end of prophase I, as in mitotic prophase, the nucleolus disappears, the nuclear membrane fragments, and a spindle apparatus is constructed.

Homologous pairs of chromosomes remain associated at chiasmata. It is, therefore, pairs of chromosomes that align themselves on the equator of the spindle during **metaphase I**. This contrasts with mitotic metaphase when homologous chromosomes move independently of each other; pairs of chromatids arrange themselves on the mitotic equator. During **anaphase I**, the homologous chromosomes separate and pairs of chromatids are pulled to opposite poles. Each new nucleus thus receives a haploid set of chromosomes, although each chromosome is present as a pair of chromatids (Figure 5.8). The second meiotic division separates these pairs of chromatids.

Following **telophase I**, when a nuclear envelope re-forms around each group of haploid chromosomes, cell division occurs. The resulting cells may proceed directly into **meiosis II** or there may be a rest phase, the length of which is species specific. The events of meiosis II are essentially the same as those of mitosis (Figure 5.4). Pairs of chromatids align on the equator and are pulled apart during anaphase II. However, the cells that are produced at the end of the process are haploid – the number of chromosomes has been halved compared to those present at the beginning of meiosis.

5.9 The biological significance of meiosis

Meiosis ensures constancy of chromosome number from generation to generation. Prior to fertilization it results in haploid cells so that when gametes fuse, a diploid zygote is produced. Meiosis also ensures variability among progeny because it promotes new combinations of alleles. This is a consequence of:

1. Independent assortment of chromosomes at metaphase I. This process varies which of two or more possible alleles at one locus is found with which allele at another locus on a different chromosome (discussed further in Section 5.10).

2. Swapping of pieces of homologous chromosomes during prophase I. This results in new combinations of alleles on the *same* chromosome (more on this in Section 7.3).

3. Random fertilization of gametes.

Together, these three processes achieve maximum mixing of the available genetic material. It is, however, *only* mutations that produce new variants, or alleles.

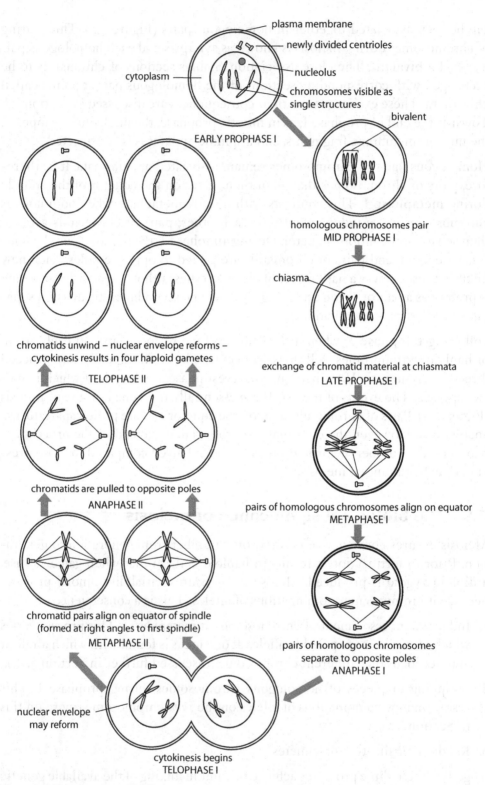

plasma membrane

newly divided centrioles

cytoplasm

nucleolus

chromosomes visible as
single structures

EARLY PROPHASE I

bivalent

homologous chromosomes pair
MID PROPHASE I

chiasma

exchange of chromatid material at chiasmata
LATE PROPHASE I

pairs of homologous chromosomes align on equator
METAPHASE I

pairs of homologous chromosomes
separate to opposite poles
ANAPHASE I

chromatids unwind – nuclear envelope reforms –
cytokinesis results in four haploid gametes
TELOPHASE II

chromatids are pulled to opposite poles
ANAPHASE II

chromatid pairs align on equator of spindle
(formed at right angles to first spindle)
METAPHASE II

nuclear envelope
may reform

cytokinesis begins
TELOPHASE I

Figure 5.8 Meiosis in an animal cell.

5.10 Revisiting Mendel's laws

The key ideas of Mendelian inheritance – that pairs of alleles segregate into separate gametes and that different pairs of alleles assort independently of each other – can be explained by the behavior of chromosomes during meiosis.

Mendel's first law (**The Law of Segregation**) states that when an organism forms gametes, only one of a pair of alleles enters each gamete. The process of meiosis explains this segregation of alleles. Consider one gene with two alleles **A** and **a**. This gene will be found at a particular locus on a chromosome. Figure 5.9 illustrates key aspects of the behavior of the relevant pair of homologous chromosomes, and therefore alleles **A** and **a**, during meiosis in a heterozygote.

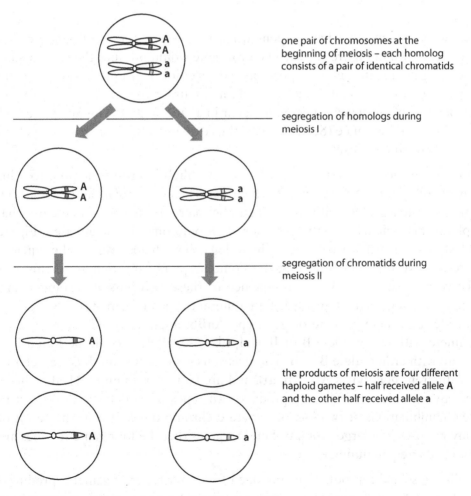

one pair of chromosomes at the beginning of meiosis – each homolog consists of a pair of identical chromatids

segregation of homologs during meiosis I

segregation of chromatids during meiosis II

the products of meiosis are four different haploid gametes – half received allele **A** and the other half received allele **a**

Figure 5.9 Meiosis segregates pairs of alleles into separate gametes.

| heterozygote parents | | Aa | | × | | Aa | |

| possible gametes | | A | a | | | A | a |

result	gametes	A	a
	A	AA dominant	Aa dominant
	a	Aa dominant	aa recessive

offspring phenotypic ratio   3 dominant : 1 recessive

Figure 5.10 Predicting the outcome of heterozygote matings.

We can next consider what happens when two heterozygotes of genotype **Aa** mate. As shown in Figure 5.9, each heterozygote would produce the two possible types of gametes in equal proportions. Fertilization is a random event. Thus, four outcomes are equally likely, which, as Figure 5.10 reminds us, results in two possible phenotypes expected to occur in a ratio of 3 : 1. This key pattern of monohybrid inheritance (Section 2.1) is thus explained by the behavior of chromosomes during meiosis.

From the results that Mendel gained from his dihybrid crosses he produced his **Law of Independent Assortment**, which states that either of a pair of alleles can segregate into a gamete with either of another pair. The key event in meiosis that explains this conclusion is the behavior of chromosomes at metaphase I. Figure 5.11 shows that when two pairs of homologs align themselves on the equator, there are two different ways that the homolog pairs can orientate themselves relative to each other. As a consequence of these two possible arrangements of the homologs at metaphase I, four different types of gametes are possible. Considering a heterozygote of genotype **AaBb**, allele **A** could segregate into a gamete with either allele **B** or **b**, and, likewise, allele **a** could be found in a gamete with either allele **B** or **b**. These various assortments produce gametes of four different genotypes **AB**, **Ab**, **aB**, and **ab**. However, Figure 5.11 also shows that any one meiosis will only produce two types of gametes with respect to allele combinations at two loci on separate chromosomes. It is as the result of many meioses in a large group of heterozygotes that the four gametic types are produced in equal numbers.

As, during sexual reproduction, any one of the four types of gametes produced by a female double heterozygote can be fertilized by any one of the four types of

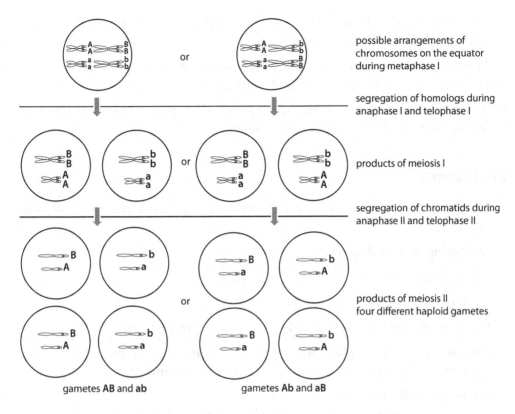

Figure 5.11 The two possible outcomes of meiosis in a heterozygote of genotype **AaBb**.

gametes produced by a male double heterozygote, there are $4 \times 4 = 16$ possible combinations of gametes. It may be remembered that these 16 possible fertilizations produce nine different genotypes, which fall into four different phenotypic classes with an expected frequency ratio of $9 : 3 : 3 : 1$ (Figure 4.2). Again, a key pattern of Mendelian inheritance is explained by the behavior of chromosomes during meiosis.

Summary

- Genes are found at specific loci on chromosomes.
- Chromosomes occur in homologous pairs in diploid cells and singly in haploid cells.
- The number of homologous pairs of chromosomes varies between species.
- The process of mitosis produces new nuclei with the same number and types of chromosomes as the parental chromosome.

- The process of meiosis produces nuclei with half the number of chromosomes as the parental chromosome.
- Meiosis produces nuclei with new combinations of alleles which ensure variability among the progeny of sexual reproduction.
- The behavior of chromosomes during meiosis explains Mendel's Laws of Segregation and Independent Assortment.

Problems

1. Specify whether the following events occur during mitosis, meiosis I, and/or meiosis II:

   (a) Pairing of homologous chromosomes.

   (b) Alignment of chromosomes along the equator.

   (c) Separation of sister chromatids.

   (d) Attachment of pairs of sister chromatids to spindle fibers.

2. The mosquito, *Culex pipiens*, has a diploid chromosome number of six. Draw diagrams to show the arrangement of chromosomes during:

   (a) Anaphase II of meiosis.

   (b) Metaphase of mitosis.

   (c) Metaphase I of meiosis.

3. List six important differences between mitosis and meiosis.

4. (a) What are (i) homologs and (ii) sister chromatids?

   (b) How similar to and different from each other are homologs and sister chromatids, respectively, with regard to genes and alleles?

5. Diploid cells of the fruit fly, *Drosophila melanogaster*, contain eight chromosomes. How many different random arrangements of homologs could occur during metaphase I of meiosis?

6. In the Madagascar periwinkle, *Catharanthus roseus*, there are six chromosomes per haploid set. How many chromosomes would you expect to observe if you prepared karyotypes of the following?

   (a) A leaf cell.

   (b) The sperm nucleus of a pollen grain.

   (c) A petal cell.

7. Match events from the list below with the appropriate stage of the cell cycle [(i) G_1; (ii) S phase; (iii) G_2; (iv) M phase]:

(a) Growth period between replication and nuclear and cell division; each chromosome has two sister chromatids.

(b) Growth period between division and replication; each chromosome has one chromatid.

(c) Chromosome replication.

(d) Nucleus undergoes mitosis followed by cell division.

8. Indicate which of the cells ($2n = 6$) in the figure corresponds to the following meiotic stages:

(a) Prophase I.

(b) Metaphase II.

(c) Anaphase I.

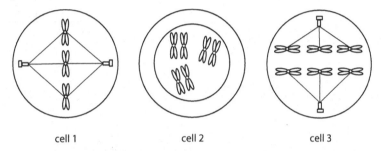

cell 1   cell 2   cell 3

9. Chimpanzees have 48 chromosomes in each somatic cell.

(a) How many chromosomes does a baby chimpanzee receive from its mother?

(b) How many autosomes and how many sex chromosomes are present in each somatic cell?

(c) How many chromosomes are present in each sperm?

10. A Shetland pony, with a diploid set of 64 chromosomes, was sharing a field with a male zebra, which has 44 chromosomes in each diploid cell. Unexpectedly, the Shetland pony gave birth to a foal. How many chromosomes would you expect in each somatic cell?

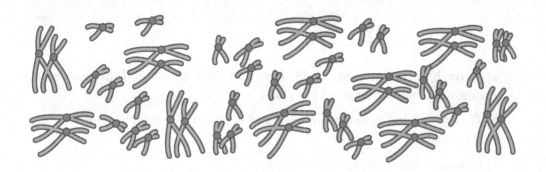

Sex Determination and Sex Linkage

The first question most people ask when a child is born is whether it is a boy or girl. The sex of the newborn child is, in fact, determined 9 months earlier, at the moment of conception. It depends upon which of two sex-determining chromosomes are present in the fertilizing sperm. Humans, as do many other species, have a chromosomal-based mechanism for determining the sex of an individual.

This chapter presents information on:

- How sex is determined in a range of different organisms

- The inheritance patterns and expression of genes located on the sex-determining chromosomes

6.1 Sex-determining chromosomes

In most higher animals and some flowering plants, the sex of an individual is determined by a particular pair of chromosomes – the **sex chromosomes**. Generally, there are two types of sex chromosomes; for example, in humans and other mammals there is an **X chromosome** and a **Y chromosome**. Any individual with two X chromosomes is female, while an individual with one X and one Y chromosome is male. The karyotype of a normal female human was shown in Figure 5.2. Chapter 5 established that the chromosomes of diploid cells are present in homologous pairs with the same genes present at the same locations on each chromosome of a pair. This remains true for an XX female, but in an XY male the sex chromosomes are non-homologous, with different genes on each chromosome. Indeed, the Y chromosome is much smaller (Figure 6.1) and carries few functional genes compared with the X chromosome.

There are, however, small segments (**pseudoautosomal segments**) at the tip of the Y chromosome that are homologous with the X chromosome. This limited homology ensures a male's sex chromosomes pair during prophase I of meiosis and so segregate into separate cells during anaphase I. Thus, one product of the

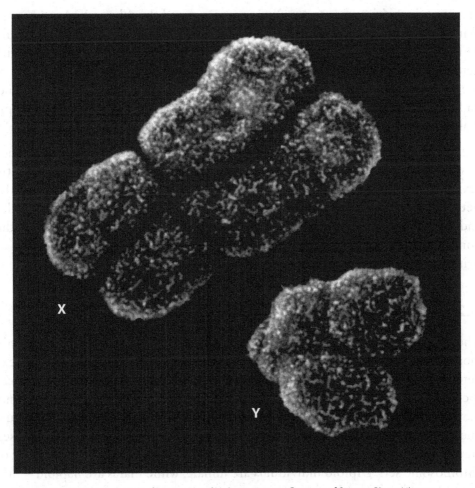

Figure 6.1 Photomicrograph of human X and Y chromosomes. Courtesy of Science Photo Library.

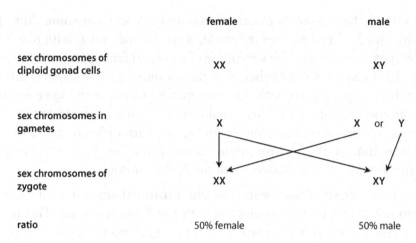

Figure 6.2 Sex determination in mammals (note that it occurs at fertilization and that there is a 50% chance of either sex being produced).

first meiotic division receives an X chromosome and the other a Y chromosome. Without this pairing mechanism for the X and Y chromosomes, gametes would be formed without, or with too many, sex chromosomes, and so problems with determination of the sex of embryos resulting from fertilization. As it is, 50% of sperm receive an X chromosome and 50% of sperm receive a Y chromosome, and we have an explanation for the (approximately) equal sex ratio among mammalian offspring (Figure 6.2).

Figure 6.2 shows that all the female gametes carry the same kind of sex chromosome. Female mammals are therefore sometimes described as the **homogametic** sex. Males produce two different types of gametes and are referred to as the **heterogametic** sex. By contrast, in birds, butterflies, moths, and some reptiles and fish, it is the female who is the heterogametic sex and the male who is the homogametic sex. To avoid confusion, the sex chromosomes in these groups of animals are called **W** and **Z** – the female is ZW and the male ZZ.

6.2   Other sex-determining mechanisms

The presence of a pair of similar or dissimilar sex chromosomes is not the only means by which the sex of an individual is determined. A variety of other sex-determining mechanisms exist throughout the living world. In some insect species, such as the grasshopper, there is only one type of sex chromosome, the X. Females are produced when two X chromosomes are present (**XX**). Just one X chromosome (**XO**) results in a male. Ants, bees, and wasps lack any sex chromosomes. Instead, the sex of an individual is determined by its number of chromosome sets. Diploid individuals are female, while males are haploid – rare examples of haploid animals!

In some invertebrates, a single locus with two alleles determines gender. For example, in mosquitoes, males are heterozygous and females homozygous recessive for a sex-determining gene. Related to this are the mating-type genes found in fungi. There are not separate male and female fungi, yet individual hyphal networks must be of different mating types to successfully fertilize. Mating type is determined by alleles at a single locus.

There are even a few examples, scattered around the animal and plant kingdoms, where the sex of a developing individual is determined by environmental factors. Reptile eggs exposed to high temperatures produce mostly males in some species and females in others (Figure 6.3). So might the dinosaurs have become extinct because the temperature rose and only one sex was produced?! The sex of some fish is determined by social dominance, while that of various marine worms and gastropods depends on the substrate upon which larvae settle. Larvae of the slipper limpet, *Crepidula fornicata*, develop into males if near a developing female and into females if further away from a female.

Certain plant species produce male or female flowers depending upon day length, although usually sex determination of flowering plants has a chromosomal basis. About 90% of flowering plants are hermaphrodite, so each flower

Figure 6.3 Female American alligator on her nest mound. During a critical period 20–35 days after hatching, if the egg incubation temperature is 30°C or below, females are produced, while 34°C or above prompts develop of males. Between 30°C and 34°C mixed sex ratios occur.

has male and female reproductive organs. Of the remaining 10%, some species are **monoecious**, bearing separate male and female flowers on the same plant, while other species are **dioecious**, having separate male and female plants. If you have ever felt frustrated at Christmas by the holly tree in your garden that never bears any bright red berries, and so is useless for decorating the house, now you know the reason – it is because it is a male tree! The holly, *Ilex aquifolium*, is an example of a dioecious plant. Among monoecious and dioecious plant species, sex-determining mechanisms vary, as they do with animals and, like animals, they seem to promote sexual development along one pathway or another. All flowers seem to have the potential to be hermaphrodite. Depending upon the sex chromosomes present, full development of either female or male flower parts is promoted and of the other repressed. Thus, if an XY system is operating, the Y chromosome carries genes that are needed for development of male flower parts while suppressing development of female parts.

6.3 The Y chromosome and sex determination in humans

Although the Y chromosome is much smaller and carries many fewer functional genes than the X chromosome, it is the presence of a Y chromosome that is the trigger for development of the male phenotype. The crucial role of the Y chromosome was first suggested by the phenotypes of individuals with unusual sex chromosome karyotypes.

Occasionally, segregation of the sex chromosomes does not occur properly during anaphase I or II of meiosis (more in Chapter 8). Gametes result that lack or contain extra sex chromosomes, and likewise the zygotes produced at fertilization. For example, if the sperm that fuses with the egg lacks a sex chromosome, the resulting zygote will be **XO**. The individual is female (Turner syndrome, see Section 8.3). If a normal ovum (X) is fertilized by an XY sperm, the zygote will be XXY and develops into a male (Klinefelter syndrome, see Section 8.4). These two karyotypes and associated phenotypes suggest that the presence of a Y chromosome is the switch needed for development along a male pathway because:

- The absence of a Y chromosome (XO) results in a female.

- The presence of a Y chromosome, regardless of the number of X chromosomes (XY or XXY) promotes the development of a male.

In 1991, the key male-determining gene (*S*ex determining *R*egion *Y*) on the Y chromosome was identified. During the early stages of embryological development the gonads are capable of developing into either ovaries or testes. The embryo's hormonal environment decides their fate. A functioning *SRY* gene triggers the production of testosterone and the gonads develop into testes.

Femaleness is, therefore, the default pathway! Very rarely, "sex-reversed" individuals are identified (i.e. **XX males** and **XY females**). In such cases one of the X chromosomes of an XX male invariably has an extra segment, derived from the Y chromosome, containing the *SRY* gene. XY females generally lack the same region on their Y chromosome.

Sex chromosomes, therefore, determine the sex of an individual. When the XY system is operating, an active *SRY* gene on the Y chromosome is crucial in triggering development of a male phenotype. Sex chromosomes do not, however, carry all the genes responsible for sexual characteristics – they are scattered throughout the autosomes. Likewise, genes determining non-sexual characteristics are found on sex chromosomes, in particular on the X chromosome (Section 6.4). As females possess two copies of an X chromosome and males only one set, this results in genes on the X chromosome having a different pattern of expression in the two sexes. The recessive phenotype is much more common in males.

6.4   The expression of X-linked genes

When considering **X-linked genes** (i.e. genes located on the X chromosome), the possible genotypes are different for males and females. Males cannot be homozygous or heterozygous for alleles of an X-linked gene as they can only possess one allele. Males are instead **hemizygous**. Figure 6.4 considers the possible

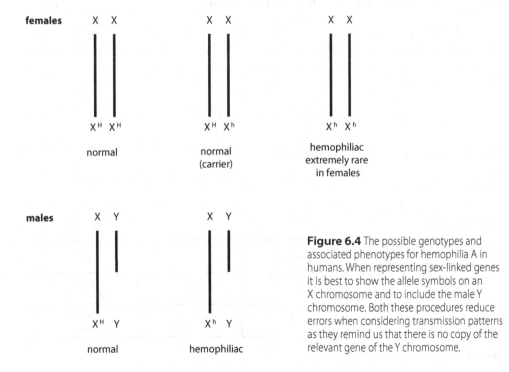

Figure 6.4 The possible genotypes and associated phenotypes for hemophilia A in humans. When representing sex-linked genes it is best to show the allele symbols on an X chromosome and to include the male Y chromosome. Both these procedures reduce errors when considering transmission patterns as they remind us that there is no copy of the relevant gene of the Y chromosome.

genotypes produced by alleles of one of the many genes found on the human X chromosome – the gene that codes for production of clotting factor VIII. The gene possesses two main alleles: the dominant **H** allele produces normal clotting factor, while the recessive allele **h** leads to a lack of this crucial protein. The blood of anyone who is homozygous or hemizygous for the recessive allele fails to clot. They suffer from the disease hemophilia.

6.5   Representing crosses involving X-linked genes

We can use genetic diagrams to show how X-linked genes are inherited in the same way as they were used for investigating patterns of inheritance of genes on autosomes. Figure 6.5 shows how a hemophiliac child can be born to parents, neither of whom express the condition.

Thus, if the woman is heterozygous, or a **carrier**, for the recessive allele, then, as shown in Figure 6.5, each time the couple conceive a child they have a 1 in 4 chance of their child being a hemophiliac. Expressed another way, if the couple have a daughter her blood will definitely clot normally, but if they have a son he has a 1 in 2, or 50%, chance of being a hemophiliac. This example illustrates the greater expression of X-linked conditions in males – a phenomenon discussed further in the next two sections.

6.6   Sex-linked inheritance patterns

In the examples of monohybrid and dihybrid inheritance discussed in Chapters 2–4, reciprocal crosses gave the same results. For example, when the inheritance of red- and yellow-fruiting tomatoes was discussed in Chapter 2, it did

parental phenotypes	normal man		×	normal woman	
parental genotypes	$X^H Y$			$X^H X^h$	
possible gametes	X^H Y			X^H X^h	
possible genotypes and phenotypes among children	gametes	X^H		X^h	
	X^H	$X^H X^H$ normal		$X^H X^h$ normal (carrier)	
	Y	$X^H Y$ normal		$X^h Y$ hemophiliac	

Figure 6.5 Investigating the inheritance of hemophilia A.

	cross 1			cross 2		
parents	female red eyed	×	male white eyed	female white eyed	×	male red eyed
F₁ result		all red-eyed		females: all red eyed males: all white eyed		
F₂ result	females: all red eyed males: 50% red eyed 50% white eyed			females: 50% red eyed 50% white eyed males: 50% red eyed 50% white eyed		

Figure 6.6 Investigating the inheritance of eye color in *D. melanogaster*: the results of a reciprocal crosses.

not matter which parent showed which phenotype. By contrast, when sex-linked characters are considered, reciprocal crosses give different results. Consider, for example, reciprocal crosses investigating the inheritance of eye color in the fruit fly, *Drosophila melanogaster* (Figure 6.6). Normally, the eyes of this fly are brick-red. White eyes sometimes occur because of a mutation in a gene that codes for a transmembrane transporter protein. The changed protein no longer transports the red pigment precursors into the eye pigment cells, so eyes appear white.

The F_1 result of cross 1 (Figure 6.6) clearly shows that red eyes are dominant to white eyes. The fact that the F_1 and F_2 results are different depending upon which parent was white-eyed and which red-eyed indicates that the encoding gene is sex linked. Figure 6.7 shows the genotypes of the flies, thus illustrating some other features associated with sex-linked genes:

- Heterozygous females transmit their X-linked recessive allele to approximately half of their daughters and half of their sons. Expression only occurs in their sons (see crosses 1 and 2).

- Males that inherit an X-linked recessive allele exhibit that trait as their Y chromosome has no counterpart. In contrast, females need two copies of a recessive allele to express the recessive trait (cross 2).

Greater expression of the recessive condition is therefore observed in male flies. The fruit fly is another example of a species showing the XY system of sex determination. In the ZW system, when females are the heterogametic sex, the pattern of expression is reversed compared with the XY system. Greater expression of a recessive Z-linked allele occurs in females.

	cross 1			cross 2		
parents	female red eyed $X^R X^R$	×	male white eyed $X^r Y$	female white eyed $X^r X^r$	×	male red eyed $X^R Y$
F_1 result	$X^R X^r$		$X^R Y$	$X^R X^r$		$X^r Y$

all red eyed

females: all red eyed
males: all white eyed

F_2 result

females: 50% $X^R X^R$ (red eyed)
        50% $X^R X^r$ (red eyed)
males: 50% $X^R Y$ (red eyed)
       50% $X^r Y$ (white eyed)

females: 50% $X^R X^r$ (red eyed)
        50% $X^r X^r$ (white eyed)
males: 50% $X^R Y$ (red eyed)
       50% $X^r Y$ (white eyed)

Figure 6.7 Investigating the inheritance of eye color in *D. melanogaster*: genotypes of flies in Figure 6.6.

6.7 Pedigree analysis of human X-linked recessive inheritance

Humans, as discussed earlier in the chapter, use the XY system of sex determination. Thus, expression of an X-linked condition is more common in males. If a human pedigree shows a pattern of greater male expression for a trait, it can immediately be suspected that the relevant gene lies on the X chromosome. Just as we identified rules for inferring autosomal patterns (Section 3.9), we can now identify useful clues for detecting X-linked inheritance in human pedigrees.

The most important indicator that a condition might be controlled by an X-linked gene is when many more males than females show the relevant phenotype. Two other clues can then confirm our suspicions:

- None of the children of an affected male will show the condition, but it may be seen again among his grandchildren (Figure 6.8). This is because only a female can inherit the recessive allele from an affected male. She will be heterozygous and so not express the condition, but she could then pass the recessive allele to any of her sons who would, of course, express it.

- There will be no inheritance of the condition through the male side of the family of an affected male. This lack of male-to-male transmission is because sons obtain their Y chromosome from their father and their X from their mother. Males do not normally inherit their father's X chromosome, with a possible recessive allele.

Perhaps the most familiar example of an X-linked condition in humans is hemophilia (discussed in Section 6.5). Famously, the recessive allele arose

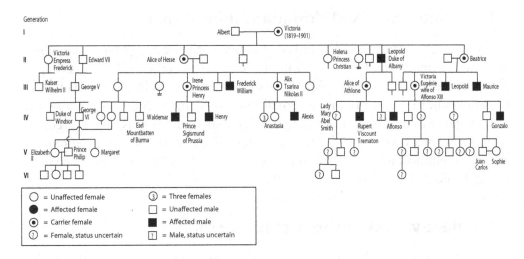

Figure 6.8 Pedigree of Queen Victoria and her descendants showing the spread of the hemophilia allele into several European royal families.

spontaneously within the reproductive cells of Queen Victoria, or one of her parents, and spread through intermarriage to many European royal families. The royal pedigree illustrated in Figure 6.8 illustrates the key features of X-linked recessive inheritance outlined above. Some other examples of human X-linked recessive conditions are shown in Table 6.1.

TABLE 6.1 A selection of human X-linked recessive conditions

CONDITION	BRIEF DESCRIPTION	INCIDENCE IN MALES
Red–green color blindness	Failure to distinguish red and green, due to lack of photopigments in color-perceiving retinal cones	1 in 125
Duchenne muscular dystrophy	Gradual wasting and atrophy of muscles from age of onset, before 6 years of age, until death around age 20; caused by lack of muscle protein dystophin	1 in 3500
Fragile X syndrome	Mental retardation and distinctive physical traits (e.g. elongated face, large ears, and large testes); caused by multiple CGG repeats in the *FMR1* gene	1 in 2000
Lesch–Nyhan syndrome	Uncontrolled movement and self-destructive biting and scratching	1 in 10,000
Testicular feminization syndrome	XY males who develop as females due to a lack of androgen receptors on cell membranes	1 in 65,000

6.8   Human X-linked dominant inheritance

There are few examples of X-linked dominant inheritance. One dominantly inherited X-linked condition is vitamin D-resistant rickets (also known as hypophosphatemia because blood phosphorus levels are low in individuals suffering from this condition). Any dominant condition, autosomal or X linked, is expressed in each generation. The indication that a dominant condition is X linked comes again from a sex bias in its inheritance – affected males will be observed to only pass the condition to their daughters, while affected females will produce both affected sons and daughters (Figure 6.9).

6.9   Inactivation of the X chromosome

As males have just one X chromosome in their genome, compared with females who have two, it would appear that males are at a potential disadvantage in having a reduced amount of an X-linked gene product. Research in the early 1960s showed that females have, in fact, opted for the male pattern of gene expression! In every female cell only one of the X chromosomes is ever functional. Early in development one of the X chromosomes in each cell becomes inactivated. The inactive X chromosome becomes highly condensed and is visible microscopically as a dark-staining spot, the **Barr body**, close to the nuclear membrane. In each embryological cell which of the two X chromosomes is inactivated is a random event. However, all subsequent mitoses from that first inactivated cell produces cells with the same X chromosome inactivated (Figure 6.10). As a result of X chromosome inactivation, the adult female body is a **mosaic** of cells with a different X chromosome inactivated in each block. The two alleles of a heterozygote

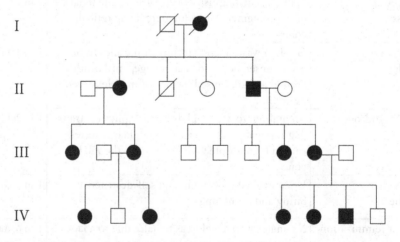

Figure 6.9 Pedigree of a family with vitamin D-resistant rickets.

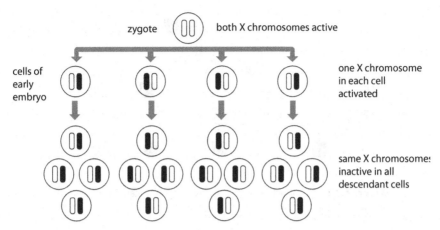

Figure 6.10 X chromosome inactivation. Early in development one X chromosome is deactivated (shown darkened). The same X chromosome is deactivated in all descendent cells.

will, therefore, be expressed in patches. This explains the mosaic pattern color of a calico (or tortoiseshell) cat or the patchy lack of sweat glands in the rare human condition of anhidrotic ectodermal dysplasia (Figure 6.11).

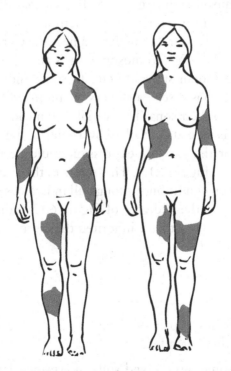

Figure 6.11 The expression of anhidrotic ectodermal dysplasia in two identical twins, heterozygous for the condition. The shaded regions represent absence of sweat glands and the different mosaic pattern in the sisters results from X inactivation.

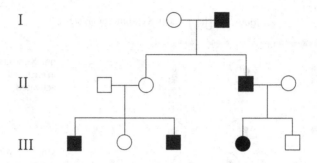

Figure 6.12 An idealized pedigree demonstrating the inheritance of an incompletely penetrant dominant disease allele.

6.10   Y-linked inheritance

Genes on the Y chromosome will only be inherited by males. Thus, we might expect a few genes that are male specific, i.e. that are only expressed in males and show male-to-male transmission (e.g. the *SRY* gene that determines maleness, discussed in Section 6.3). Hairy ear rims are sometimes quoted as a Y-linked condition. It is a condition only found in males – a necessary criterion of a Y-linked condition as only males have Y chromosomes. It is an extremely rare condition most often seen in the Indian subcontinent. In some Indian families expressing this trait every male possesses hairy ears, while in others only some males transmit the condition to their sons. This variability means we have to question whether hairy ear rims is indeed controlled by a gene on the Y chromosome. One possible explanation for the variable expression could be that the gene shows **incomplete penetrance** (Figure 6.12). This refers to the situation when individuals fail to express a given condition, although all indicators are that they possess an expressing genotype. In the pedigree of Figure 6.12, individual II-2 does not express the disease, yet she must have inherited the disease allele to pass it on to her two children III-1 and III-3.

Summary

- In many diploid species, a pair of chromosomes – the sex chromosomes – determine an individual's gender.

- In higher animals, some insects, and some flowering plants, the female possesses two X chromosomes in each somatic cell, and is the homogametic sex. The male has an X and a Y chromosome in each cell, and is the heterogametic sex.

- The presence of a functional *SRY* gene on a Y chromosome triggers development of a male phenotype; otherwise, an individual develops as a female.

- In birds, butterflies, and moths, the female is the heterogametic sex.

- Many genes determining non-sexual characteristics are found on the sex chromosomes and are described as sex linked.

- With respect to the XY system, because males have only one X chromosome, all alleles, recessive as well as dominant, are expressed. As a consequence, different patterns of inheritance are observed in males and females for X-linked traits.

- Although a female has two X chromosomes, only one is expressed in each cell. The other X chromosome is inactive and exists in an inactive state known as a Barr body.

Problems

1. If a woman heterozygous for a recessive sex-linked allele marries a man who does not show the trait, what is the probability that:

    (a) The recessive allele will be passed on to any child the couple might have?

    (b) The couple might have a child who shows the trait?

    (c) Any of their sons would show the trait?

2. What would be the most likely genotypes of parents whose daughters were all normal sighted and whose sons were all color blind?

3. A woman whose father suffered from hemophilia A is married to a normal man who thinks that his grandfather might have had the condition. What are the couple's chances of having affected children if the man's grandfather did in fact suffer from hemophilia?

4. Black fur in cats is a result of an X-linked allele, **B**, while the alternative allele, **Y**, results in ginger fur. Heterozygotes have a patchy black and ginger coat, referred to as tortoiseshell. What kittens might result from a mating of a black male and tortoiseshell female?

5. What advice might a genetic counselor offer to a woman whose mother's brother died, aged 19, of the X-linked recessive condition, Duchenne muscular dystrophy? Her husband is in full health and there is no history of the condition in his family. She is concerned about the possibility of any children they might have suffering from the disease.

6. Among a litter of four Dalmatian puppies, what is the probability that:

    (a) Two are males and two are females?

(b) All four are females?

(c) At least one puppy is a female?

7. Females have been observed with four Barr bodies in their cells. How many X chromosomes are present in each somatic cell?

8. If a rare genetic disease in horses is inherited on the basis of an X-linked dominant allele, would an affected mare or an affected stallion always have affected female foals?

When answering Questions 9 and 10 remember that the female is the heterogametic sex.

9. Several pairs of gray cockatiels were crossed. In all cases the nests contained both white- and gray-feathered chicks, but all the white birds were female. When white-feathered birds were crossed, all the resulting chicks were white. What is the probable basis of inheritance of feather color in these birds?

10. It is generally impossible to sex chicks morphologically. However, commercial chicken breeders have devised a way, in some breeds, based on plumage color. Barred plumage (**B**) is dominant over non-barred (**b**) or solid color. This difference is encoded by a gene located on the Z chromosome. What must be the genotypes of the male and female parents in order to be able to sex the chickens soon after hatching?

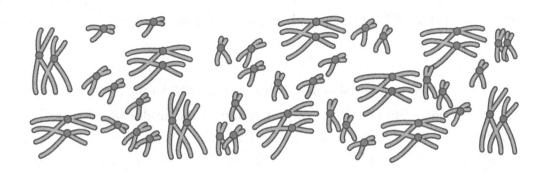

Linkage and Chromosome Mapping

It is estimated that the human genome possesses approximately 21,000 genes. These are distributed between just 23 different chromosomes. A major goal of genetics during the early twenty-first century is to **map** (i.e. determine) the precise chromosomal position of all the human genes, as well as those of many other species. Mapping genes has become a sophisticated molecular process. However, traditional Mendelian analysis has been, and continues to be, a useful tool in the initial stages; the results of controlled genetic crosses are used to produce preliminary or **outline maps**. These show which of an organism's genes are gathered together on a particular chromosome and their approximate positions relative to each other. Molecular analysis is then used to precisely locate these genes.

This chapter describes the contribution made to chromosomal mapping by Mendelian genetics, in particular how:

- Distinctive dihybrid phenotypic ratios can indicate when two genes are located on the same chromosome.

- These ratios can be used to map genes relative to each other.
- The events of prophase I of meiosis results in the production of these ratios.

7.1   Linkage and recombination

In the late 1900s, Thomas Morgan (Box 7.1) was studying the inheritance of vestigial wings and purple eye color in the fruit fly, *Drosophila melanogaster*. The results of one of his crosses is shown in Figure 7.1. Morgan had used the standard Mendelian technique of crossing male and female flies that were pure breeding for the two pairs of contrasting characters (i.e. red-eyed, normal winged with purple-eyed, vestigial winged) to generate a F_1 generation (Figure 7.1). Among the various crosses that he then performed was a backcross of the F_1 flies to the double-mutant parent. Morgan obtained an unexpected result. He had expected to obtain all possible pairs of characters (i.e. red-eyed, normal; red-eyed, vestigial winged; purple-eyed, normal winged; and purple-eyed, vestigial winged) in equal proportions (this was a test cross, see Section 4.2). Instead, the two parental phenotypic classes (red, normal and purple, vestigial) predominated (Figure 7.1), while the other two phenotypic classes, the recombined characters (red-eyed, vestigial winged and purple-eyed, normal winged) were underrepresented. The wing and eye genes were not being inherited independently.

Morgan had discovered genetic linkage. The wing length and purple eye gene are carried on the same chromosome – they are **linked**. Alleles of these two genes are inherited together unless a physical exchange of chromosome segments occurs

parental cross	pure breeding red eyes, normal wings		×	pure breeding purple eyes, vestigial wings	
F₁ result		all red eyes, normal wings			
test cross	F₁ red eyes, normal wings		×	pure breeding purple eyes, vestigial wings	
test cross result	1339 red eyes, normal wings	151 red eyes, vestigial wings	154 purple eyes, normal wings	1195 purple eyes, vestigial wings	
ratio	8 :	1 :	1 :	8	

Figure 7.1 Investigating the inheritance of eye color and wing presence in *D. melanogaster*.

BOX 7.1 THOMAS MORGAN AND THE FRUIT FLY

Thomas Hunt Morgan, an American, was one of the key geneticists of the first half of the twentieth century. Inspired by the rediscovery of Mendel's work, he began his own breeding experiments with the fruit fly, *D. melanogaster*. He soon realized he had chanced upon an ideal organism for heredity work. It was extremely prolific, had a short generation time (10–14 days), and possessed many easily identifiable morphological variants (e.g. different eye colors and shapes). Morgan's first breeding experiment, in 1910, was between a normal red-eyed fly and a newly discovered white-eyed mutant. The results convinced Morgan that genes are carried on chromosomes. He and his students, in particular Calvin Bridges, Hermann Muller, and Arthur Sturtevant, followed this initial discovery with a decade of crosses that developed the methods for mapping genes on chromosomes and firmly established the fruit fly as a model organism for genetic study. Over the decades, the fruit fly has contributed enormously to many aspects of developmental and molecular biology. Its genome was sequenced in 2000 and valuable insights into the pathology of complex human diseases, such as Parkinson's disease, have been provided by working with transgenic flies. In 1933, Morgan was awarded the Nobel Prize for Physiology or Medicine.

(A)   (B)

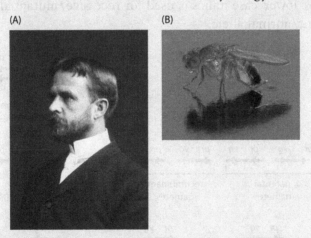

Box 7.1 Figure 1 (A) Thomas Morgan. Courtesy of André Karwath under CC BY-SA 2.5 license. (B) *Drosophila melanogaster*.

during prophase I of meiosis. This produces new combinations of the alleles. The exchange of chromosomal segments is termed **recombination**.

7.2 Representing dihybrid crosses involving linked genes

Hitherto, genotypes have been represented solely by letters (e.g. the double heterozygote as **AaBb**). When considering crosses involving linked genes it is

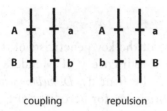

Figure 7.2 Representing linked genes. A double heterozygote (**AaBb**) is shown with alleles coupled or in repulsion.

coupling   repulsion

generally better to indicate the genotype using a summary chromosome. This serves to remind us that we are dealing with linked genes. Thus, if genes **A** and **B** are located on the same chromosome, the double heterozygote would be represented as shown in Figure 7.2. This figure also shows that two arrangements of the alleles are possible for a double heterozygote. When **coupled**, the two dominant alleles are linked on one chromosome (of a homologous pair) and the two recessive alleles are on the other. **Repulsion** indicates a dominant and recessive allele on each chromosome.

Figure 7.3 shows Morgan's test cross between the double heterozygote of Figure 7.1 and homozygous recessive purple-eyed, vestigial winged flies represented in "linkage format." Note also the different format for representing alleles of *Drosophila* genes: lower-case italics is used for recessive/mutant alleles and a '+' for the dominant/normal allele.

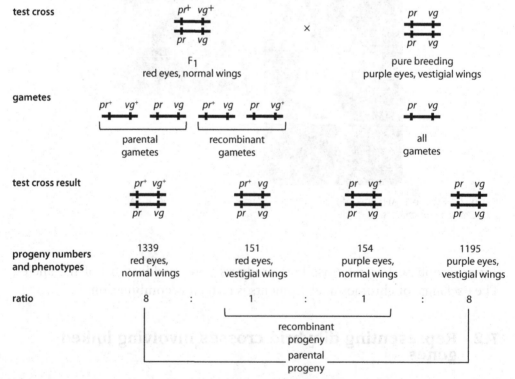

Figure 7.3 Representing Morgan's test cross in linkage format. Allele symbols: pr^+ = red eyes, pr = purple eyes, vg^+ = normal wings, and vg = vestigial wings. A superscript "+" denotes the dominant allele.

7.3 Explaining Morgan's results

The majority (89%) of Morgan's flies showed parental phenotypes (i.e. red eyes, normal wings and purple eyes, vestigial wings). This correlates with most of the gametes being parental types (pr^+ vg^+ and pr vg), and only a few being non-parental types (pr^+ vg and pr vg^+). The pr^+ vg^+ and pr vg allele pairs are physically linked on *D. melanogaster* chromosome II, and new allelic combinations of these two linked genes only occur if the process of **crossing over** between these two alleles takes place during prophase I of meiosis.

When homologous chromosomes pair during prophase I (look again at Figure 5.8) there is an opportunity for chromosomes to exchange segments. This results in new, **non-parental** or **recombinant** combinations of alleles. This exchange is called **crossing over**, because genes from one homolog have "crossed over" on to its partner. It is achieved by the **breakage** and **rejoining** of chromatids. The crossover point is known as a **chiasma** (chiasmata, plural). Figure 7.4 shows how crossing over results in the recombinant (pr^+ vg and pr vg^+) gametes in our *Drosophila* example. Note that it is non-sister chromatids that exchange segments.

Large numbers of cells may be undergoing meiosis in the reproductive organs of animals and plants. In male gonads, especially, millions of cells are generally dividing. Chiasmata occur randomly along the length of a chromosome. Indeed, several chiasmata generally form between each pair of homologous chromosomes during meiosis, illustrated in the photomicrograph of Figure 7.5. However, in only a proportion of those cells undergoing meiosis will a chiasma occur between the two loci being investigated in any specific cross. The frequency with which chiasmata form between any two genes depends upon the distance between them – fewer chiasmata form between closely linked genes.

When a chiasma occurs between two genes, **recombinant** gametes result (Figure 7.4). The number of recombinant gametes relative to the number of parental gametes will vary according to the positions of the genes under investigation. The

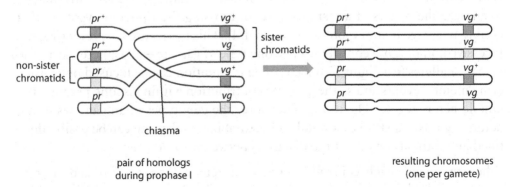

Figure 7.4 Gametes produced by the F_1 heterozygote pr^+/pr vg^+/vg when crossing over occurs between the eye color and wing presence genes.

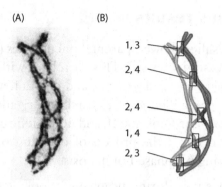

(A)   (B)

1, 3

2, 4

2, 4

1, 4

2, 3

Figure 7.5 Chiasmata. (A) Photomicrograph of chromosome 5 of the grasshopper, *Chorthippus parallelus*, during prophase I of meiosis. Five chiasmata have formed. (B) Interpretative diagram. The numbers indicate which chromatids have formed each chiasma. Note that they have always formed between non-sister chromatids.

further apart two loci are, the more recombinant gametes are produced, because of the greater opportunity for chiasmata to form between them. The result is a greater number of offspring of a cross showing recombinant phenotypes. The frequency of recombinant individuals has become a useful tool in estimating relative distances between gene loci and thus producing maps of chromosomes showing relative positions of the genes. Before explaining the principles of such **chromosome mapping**, it is first necessary to discuss the types of genetic crosses that are used.

7.4   The use of the test cross in linkage studies

Previously, when discussing dihybrid inheritance, emphasis was placed on analyzing the phenotypes of the F_2 progeny resulting from selfed or interbred F_1 heterozygotes and comparing the relative proportions of these different phenotypes with the "classical" $9 : 3 : 3 : 1$ ratio. In linkage studies, it is more informative to analyze the results of a **test** cross (i.e. of crossing the F_1 heterozygote with a doubly recessive homozygote, as Morgan did). This is because, in a test cross, the recessive homozygote contributes just one type of gamete – all gametes carry the recessive alleles (e.g. Figure 7.6). This means attention can be focused on considering meiotic events and so the gametic contribution within just one parent – the double heterozygote. The test cross exposes the genotypes of all gametes of the heterozygote for analysis. Parental and recombinant offspring can be easily identified and estimates made of the distances between two loci.

When two genes each controlling a separate character are *not* linked, a cross between a double heterozygote and the recessive homozygote produces four phenotypes among the progeny, each type present in (approximately) equal

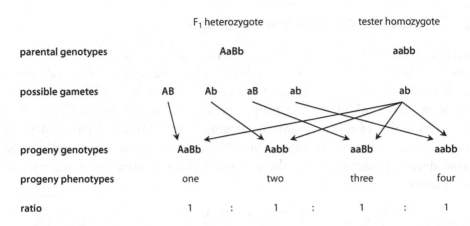

Figure 7.6 Genotypes and phenotypes of a test cross involving two unlinked genes.

numbers (Figure 7.6). If, however, the two genes under investigation are linked, a test cross produces major deviations from this 1 : 1 : 1 : 1 phenotypic ratio. Four phenotypic classes are still obtained among the progeny, but parental phenotypes greatly outnumber recombinant ones, as Morgan found. His red-eyed, normal winged, red-eyed, vestigial winged, purple-eyed, normal winged and purple-eyed, vestigial winged flies were in an approximate ratio of 7 : 1 : 1 : 7. In another cross, flies heterozygous for a gene (*scute*) inhibiting growth of thoracic bristles and a second gene (*echinus*) causing a roughened eye surface were crossed with flies recessive for these two genes. This cross produced a ratio of 15 : 1 : 1 : 15 among the four progeny classes. From the size of the deviation from the unlinked 1 : 1 : 1 : 1 ratio, conclusions can be drawn about the relative distance apart of the two genes. The greater the deviation, the closer the two genes are together; there is less opportunity for crossing over to occur between the two genes to generate recombinant gametes and so recombinant offspring.

It is possible for there to be equal frequencies of the four possible phenotypes among the progeny of a test cross if two genes are linked, but at opposite ends of the same chromosome. Being so far apart, a chiasma forms between two such genes during every meiosis and so equal numbers of all four possible gametes are produced, and segregation patterns typical of unlinked genes are obtained. Such is the situation for some of the genes controlling the pea traits that Mendel studied. Retrospectively, we know that Mendel was fortunate in the outcome of his experiments investigating dihybrid inheritance. The inheritance patterns that Mendel observed in these experiments indicated that the genes controlling each of the seven pea traits that he chose to study were inherited independently of each other. These results led him to propose his **Law of Independent Assortment**. However, some of the traits Mendel studied are controlled by genes located on the same chromosome, albeit widely separated.

7.5 Producing chromosomal maps

When the results of a test cross (few recombinant progeny and large numbers of parental progeny) indicate that two genes are linked, they can also be used to produce an estimate of the distance between the two genes. Returning to Morgan's cross investigating inheritance of purple eyes and vestigial wings (Figure 7.1). We use the number of recombinants to calculate a map distance. In total, 305 flies (those possessing red eyes, vestigial wings or purple eyes, normal wings) showed a recombinant phenotype. From these numbers we calculate the percentage of recombinants:

$$\text{Recombinants (\%)} = \frac{\text{number of recombinants}}{\text{total number of progeny}} = \frac{305}{2839} = 0.11 \text{ or } 11\%$$

This figure for the percentage of recombinants is directly convertible into a distance expressed in **map units** or **centimorgans, cM** (named after Thomas Morgan), where **1 cM** is defined as the distance between two genes that produces a recombination frequency of 1% in test crosses with a double heterozygote. Thus, the distance between the eye color and wing shape gene is 11 cM. This distance has a physical reality, as 1 cM is approximately equivalent to a segment of DNA of 10^6 nucleotide pairs.

The calculation described in the previous paragraph has given us a distance between two genes. Generally, when mapping, we are keen to know the relative positions of a number of genes. This can be achieved if we perform a number of crosses involving different gene pairs. The results generate a set of gene distances from which we deduce the correct gene order. For example, suppose we wanted to map genes **A**, **B**, **C**, and **D** relative to each other, and suppose pairwise test crosses generated the results shown in Table 7.1.

A gene order can be deduced from the map distances by a trial and error process (Figure 7.7).

It is impossible to know in this present example whether genes **B**, **C**, and **D** lie to the right or left of gene **A**, but further crosses with other genes could answer

TABLE 7.1 Map distances between genes **A**, **B**, **C**, and **D** calculated from a series of pairwise test crosses

GENES INVOLVED IN THE TEST CROSS	DISTANCE BETWEEN GENES (cM)
A/a × B/b	3
B/b × C/c	2
A/a × C/c	4.5
A/a × D/d	12.5
C/c × D/d	9

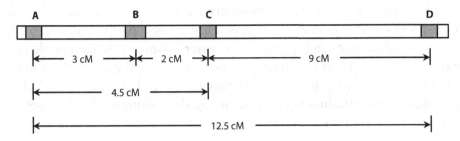

Figure 7.7 Determination of the relative position of genes **A**, **B**, **C**, and **D**, using data from Table 7.1.

this question. It should also be noticed that the map is not additive. We might have expected genes **A** and **D** to be 14 cM apart (the sum of the individual distances between genes **A** and **B**, **B** and **C**, and **C** and **D**), yet the results of the cross involving gene pair **A** and **D** suggest that they are only 12.5 cM away from each other. We can attribute this apparent discrepancy to the formation of **double crossovers**. Estimating map distances relies on the occurrence of a single crossover event between relevant genes to generate recombinant chromosomes. However, if two genes are far apart on a chromosome, two chiasmata may form between the genes of interest. As a consequence. the alleles remain in their original relationships, no recombinants are produced (Figure 7.8), and the distance between the two genes is underestimated.

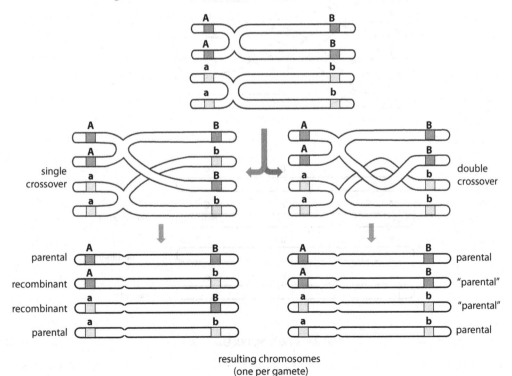

resulting chromosomes
(one per gamete)

Figure 7.8 The consequence of a single or double crossover between genes **A** and **B**.

To eliminate the effect of double crossovers artificially deflating map distances, a **three-point cross** can be made; that is, a test cross is performed with a triple heterozygote (**AaBbCc**) and the recessive tester **aabbcc**. While a double crossover between genes **A** and **C** still returns the first and third pair of alleles to their original linkage position, the middle pair are exchanged (see Figure 7.9). A rare class of **double recombinants** is found among the offspring, which is useful for map distance calculations.

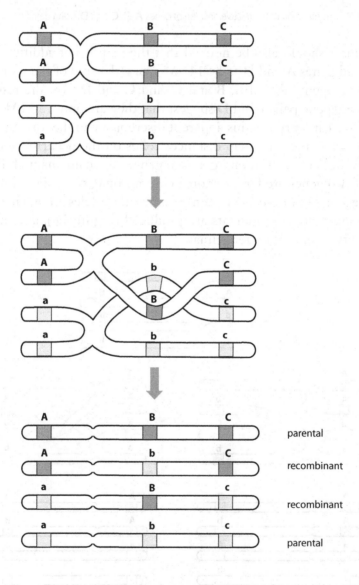

CHROMOSOMES PRODUCED

(one per gamete)

Figure 7.9 The consequences of a double crossover in a three-point cross.

7.6   Using a trihybrid test cross to map genes

In a trihybrid test cross a triple heterozygote (**AaBbCc**) is mated with an individual that is homozygous recessive for all three genes (**aabbcc**). Eight progeny classes are produced which fall into four categories (Table 7.2):

- Parental (non-recombinant) progeny
- Recombinant progeny resulting from a single crossover between genes **A** and **B**
- Recombinant progeny resulting from a single crossover between genes **B** and **C**
- Recombinant progeny resulting from a double crossover between genes **A** and **C**

There is a standard way of identifying the different classes of a trihybrid cross so that:

- The order of the three genes can be determined.
- The distances between the three genes can be calculated.

TABLE 7.2 The results of a trihybrid test cross: vermillion-eyed, cross-veinless-winged, cut-winged fruit flies (triple-recessive homozygote) were crossed with wild-type fruit flies (allele symbols: ct = cut wings, cv = cross veinless wings, and v = vermillion eyes)

PROGENY PHENOTYPE	PROGENY NUMBERS	PROGENY GENOTYPE	
Wild-type	592	$v^+ cv^+ ct^+ / v\ cv\ ct$	Parental[1]
Vermillion eyes, cross veinless, cut wings	580	$v\ cv\ ct / v\ cv\ ct$	Parental
Cross veinless	45	$v^+ cv\ ct^+ / v\ cv\ ct$	Single recombinant
Vermillion eyes, cut wings	40	$v\ cv^+ ct / v\ cv\ ct$	Single recombinant
Vermillion eyes	89	$v\ cv^+ ct^+ / v\ cv\ ct$	Single recombinant
Cross veinless, cut wings	94	$v^+ cv\ ct / v\ cv\ ct$	Single recombinant
Cut wings	3	$v^+ cv^+ ct / v\ cv\ ct$	Double recombinant[2]
Vermillion eyes, cross veinless	5	$v\ cv\ ct^+ / v\ cv\ ct$	Double recombinant

[1]The parental progeny (produced when no crossover occurs) are always the two classes with most progeny.

[2]The double-recombinant progeny (produced when a double crossover occurs) are always the two classes with least progeny.

This mapping analysis of a trihybrid cross can be demonstrated using the results obtained from mating fruit flies heterozygous for cut wings, vermillion eyes, and cross veinless wings with others homozygous recessive for these traits (Table 7.2).

The raw data presented in Table 7.2 gives the progeny phenotypes and numbers of each class. The order of the three genes and the distance between each can be deduced from this data. **Gene order** is worked out by:

• Comparing the phenotypes of the parental and double crossover progeny

• Noting which trait differs in its expression between parental and double crossover progeny; this is encoded by the middle gene

Considering the data in Table 7.2, parental and double crossover progeny are alike in the expression of cross veinless wings and vermillion eyes, and differ in the expression of cut wings. Thus, the gene encoding cut wings is in the middle. **Gene distances** are calculated separately from the recombination frequencies between each pair of genes. For each pair of genes below the number of recombinants includes progeny produced by single and double crossovers between the two genes. For the cross of Table 7.2:

$$\text{Recombinants (\%) between cross veinless and cut wings} = \frac{45 + 40 + 3 + 5}{1448} = 0.064 \text{ or } 6.4\%$$

$$\text{Recombinants (\%) between cut wings and vermillion eyes} = \frac{89 + 94 + 3 + 5}{1448} = 0.132 \text{ or } 13.2\%$$

A genetic map of these three genes is shown in Figure 7.10. Further crosses involving other genes on the same chromosomes are needed to determine the relative positions of the cross veinless and vermillion eyes genes; that is, to determine whether their positions are as shown in Figure 7.10 or whether vermillion eyes is to the left and cross veinless to the right of the cut wings gene.

7.7   Interference and coefficient of coincidence

Not even a three-point cross can, however, eliminate a few problematic crossover effects. Crossovers do not occur with the same frequency at all regions along a chromosome. There is a reduced chance of a crossover occurring near a centromere, at the ends of a chromosome, and in regions where there are few functional genes (heterochromatin). In many species, crossovers occur less often in males than females. Furthermore, the presence of a crossover can **interfere** with the development of another crossover close by. Interference has implications when analyzing the results of trihybrid mapping crosses. It results in double crossovers

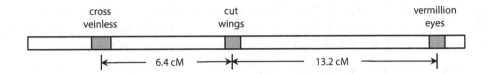

Figure 7.10 Genetic map of cross veinless, cut wings, and vermillion eyes.

being less than expected and so genetic distance being underestimated. The **coefficient of coincidence** compensates for this phenomenon:

$$\text{Coefficient of coincidence} = \frac{\text{number of observed double crossovers}}{\text{number of expected double crossovers}}$$

$$\text{Number of expected double crossovers} = \text{product of the two single crossover frequencies}$$

In the trihybrid cross of Table 7.2:

$$\text{Coefficient of coincidence} = \frac{8}{10.74} = 0.74$$

This means approximately 75% off the expected double crossovers were observed.

Interference and other factors outlined above lead to an underestimation of absolute distances between genes. However, relative distances will be the same, which, after all, is the relevant aspect of mapping. Chromosome maps are an abstract concept. They merely show the relative positions of genes. Other, physical mapping, techniques are then used to precisely locate genes on chromosomes.

7.8   A few comments on physical mapping

As the twentieth century progressed, ever more detailed chromosome maps were produced for an increasing number of species. It is, however, only in the last couple of decades that geneticists have been able to start identifying the precise location of genes on their respective chromosomes. The DNA technological revolution has made this possible.

Molecular analysis of chromosomes has shown many sites of neutral variation at the DNA level (i.e. variants with no effect on an individual's phenotype, but detectable by molecular means). These sites are generally within non-coding regions of chromosomes and can be used as **DNA markers**. Individuals can be heterozygous for these marker regions, which we can refer to as "alleles" and which are inherited in a Mendelian fashion. Crosses are made in which one of the "genes" is a DNA marker whose precise location in a genome is already known. Thus, mapping a gene, with known phenotypic effect, relative to a marker means we immediately have a definite idea of the gene's position. We can then:

1. Chop out the relevant segment of DNA.

2. Clone (i.e. make multiple copies) of this targeted segment for ease of analysis (see Chapter 13).

3. Determine the DNA sequence of the cloned segment and from its molecular character precisely locate our gene of interest.

Once the DNA sequence of a gene is known, it is possible to make predictions about the nature of the encoded protein and even its role within the cell. It is the gateway to therapies of various kinds – pharmacological and genetic.

7.9 Mapping human genes

Controlled matings are of course impossible for humans. Thus, a different method to that of analyzing the results of trihybrid and dihybrid test crosses had to be developed to map human genes. Human pedigrees are analyzed for co-inheritance of a trait (often a disease) and alleles of a genetic marker (generally a DNA marker). If a particular allele of a DNA marker and a disease allele are consistently co-expressed in many individuals in successive generations then this co-expression signals that the two genes (DNA marker and disease gene) may be linked. For example, in the pedigree in Figure 7.11 it appears that the dominant disease allele and allele 3 of the DNA marker are co-segregating. This apparent co-segregation would need to be confirmed statistically. Human geneticists calculate a **LOD score** (logarithm of **od**ds) to confirm or reject co-segregation.

If alleles of two genes consistently co-segregate, there could be two explanations for the observation:

• The two genes are linked close together on the same chromosome; *or*

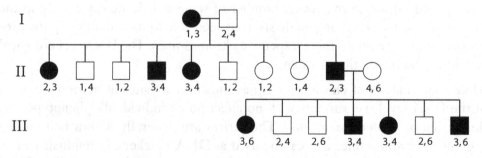

Figure 7.11 An idealized pedigree showing co-segregation of a DNA marker allele (allele 3) with the dominant disease allele (filled symbols). The numbers for each individual refer to different alleles of the DNA marker.

- The two alleles are being inherited together by chance (i.e. during meiosis the chromosome with the disease allele and the other with the particular DNA marker allele are assorting together in the same gamete).

A LOD score distinguishes between these two possibilities.

$$\text{LOD score } (Z) = \log_{10} \frac{\text{probability of observing co-segregation}}{\text{probability of observing co-segregation}}$$
$$\frac{\text{if the genes are linked}}{\text{if the genes are } not \text{ linked}}$$

We calculate a LOD score for various possible frequencies of recombinations, since we do not know how close the two genes may be, if linked (Box 7.2). A positive LOD score indicates two genes are likely to be linked. A LOD score greater than 3 is generally needed as evidence for linkage. A LOD score above 3 indicates 1000 to 1 odds that the association between the two genes observed in the pedigree did not occur by chance, but because of linkage. Negative LOD scores indicate that linkage is unlikely. A LOD score less than −2 is considered evidence to exclude linkage. LOD scores of −2 to 3 are inconclusive. They indicate genes may be linked and more data needs collecting to confirm or refute linkage. Initially, producing LOD scores can seem complicated! It is, however, useful to have a general understanding as a LOD score is a frequently quoted statistic.

BOX 7.2 CALCULATING A LOD SCORE

The pedigree below shows the co-inheritance of the dominant human condition of nail–patella syndrome (poorly developed nails and kneecaps) and ABO blood groups:

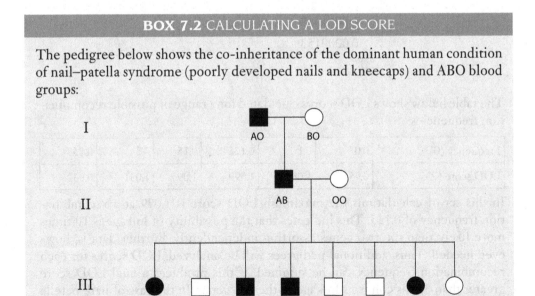

Box 7.2 Figure 1 Inheritance of nail–patella syndrome and blood group.

Blood group allele **A** appears to co-segregate with the dominant disease allele. The one exception is individual III-5, who is a possible recombinant. To assess the likelihood of linkage of these two genes, LOD scores are calculated for various possible frequencies of recombination.

$$\text{LOD score } (Z) = \log_{10} \frac{\text{probability of observing co-segregation if the genes are linked}}{\text{probability of observing co-segregation if the genes are } not \text{ linked}}$$

$$= \log_{10} \frac{(1 - \theta)^{n-r} \times \theta^r}{(0.5)^n}$$

where θ is the recombination frequency, n is the number of informative progeny, and r is the number of recombinants.

Considering generation III in the above pedigree and a recombination frequency of 0.125, as 1 in 8 of the individuals might be a recombinant:

$$\text{LOD score } (Z) = \log_{10} \frac{(1 - 0.125)^{n-r} \times (0.125)^r}{(0.5)^n}$$

$$= \log_{10} \frac{(0.875)^7 \times (0.125)}{(0.5)^8}$$

$$= \log_{10} \frac{0.0001917}{0.0000153}$$

$$= 1.099$$

The table below shows LOD scores calculated for a range of possible recombination frequencies:

Frequency (θ)	0.05	0.1	0.125	0.15	0.2	0.25
LOD score (Z)	0.951	1.088	1.099	1.09	1.031	0.932

In this set of calculations the maximum LOD score is 1.099 at a recombination frequency of 0.125. This indicates that the possibility of linkage is 10 times more likely than the two genes assorting independently. Further data is, however, needed. Thus, additional pedigrees will be analyzed. LOD scores for each recombination frequency can be summed. If this produces a final LOD score greater than 3, this confirms linkage of the two genes. In the case of nail–patella syndrome and ABO blood groups, combined LOD scores produced values greater than 3, indicating the two genes are linked, about 10 cM apart.

Summary

- Gene mapping involves determining the order of genes and distances between them.
- It involves analyzing the inheritance patterns of linked genes.
- The results of two and three point test crosses are analyzed.
- Recombinant progeny are identified.
- Distances between genes are determined from the percentage of recombinant progeny.
- Distance between genes is expressed in centimorgans (cM).
- Pedigrees are analyzed in the mapping of human genes.
- LOD scores are calculated, using pedigree data, to assess the likelihood of two genes being linked.

Problems

1. In corn snakes, two genes on the same chromosome control color. Any snake having the genotype **O-B-** is brown, **O-bb** is orange, **ooB-** is black, and **oobb** is albino. A cross between an albino and a heterozygous brown snake (**OoBb**) produced 73 brown, 24 orange, 26 black, and 69 albino offspring. How does this data indicate linkage between the two loci?

2. In the garden pea, a gene controlling sensitivity to pea mosaic virus is linked to a gene controlling pod color. Sensitivity to the virus is recessive to resistance and orange pods are recessive to green ones.

   (a) If plants true breeding for resistance and normal pods are crossed to others that produce orange pods and are sensitive to the virus, what will be the phenotype and genotype of the F_1 progeny?

   (b) If these F_1 progeny are test crossed, what are the possible phenotypes of the progeny? As the genes are linked, which phenotypes will be the most frequent?

3. A recessive allele at a locus on chromosome 5 produces an odd gait in mice referred to as waltzing. When pure-breeding albino, waltzer mice were mated with pure-breeding colored, non-waltzer mice, all the progeny had colored fur and a normal gait. However, when these F_1 mice were crossed to a waltzer, albino the following mice were obtained:

Phenotype	Number
Colored, non-waltzer	36
Colored, waltzer	7
Albino, non-waltzer	12
Albino, waltzer	38

(a) What evidence is there that the genes controlling these two traits are on the same chromosome?

(b) State the genotypes producing each of the four phenotypes in the table.

(c) Suppose an albino, waltzer mouse was crossed with one of the albino, non-waltzer mice, what progeny might you obtain?

4. Two loci are very close together on a chromosome. Suppose an individual recessive at both loci (**aabb**) was mated with a double heterozygote (**AaBb**). What phenotypes and in what proportions would you expect among the progeny if no crossing over occurred between these two loci during the meiosis giving rise to the gametes used in this mating.

5. How is it possible for two genes on the same chromosome to undergo independent assortment?

6. If 7.7% of all meioses in winter wheat result in a chromatid exchange between loci **C** and **D**, what is the map distance in centimorgans between these two genes?

7. Green is the dominant color in Pacific tree frogs. Occasionally, a rare blue frog is found. This color locus is linked to another determining the size of the frogs' toe pads. A large toe pad is recessive to a normal-sized pad. In a cross between a blue, large-padded tree frog and a green individual with normal-sized pads, which is heterozygous at both loci, the offspring consisted of 32 green, normal-padded frogs, 27 blue, large-padded frogs, nine green, large-padded frogs, and 12 blue normal-padded frogs. What is the map distance between the two loci?

8. In the fruit fly, *D. melanogaster*, alleles at one locus determine the production of normal or modified antennae (aristapedia). Another pair of alleles results in gray or ebony bodies. Flies with normal antennae and gray bodies were crossed with others that had aristapedia antennae and ebony bodies. All the flies in the resulting F_1 generation had normal antennae and gray bodies. When F_1 flies were mated with the parental flies with aristapedia antennae and ebony bodies, approximately half the progeny had aristapedia antennae and ebony bodies, and the other half normal antennae and gray bodies. In addition, in a few crosses, there were both flies with aristapedia antennae and gray bodies, and normal antennae and ebony bodies. Explain these results.

9. Two genes are on the same chromosome, 14 cM apart. Parents of genotypes **AABB** and **aabb** were crossed and the resulting heterozygous (**AaBb**) mated with recessive individuals (**aabb**).

   (a) Of the 2000 offspring produced, what numbers might you expect of the four different possible genotypes: **AaBb**, **aaBb**, **Aabb**, and **aabb**?

   (b) What would be the predicted results of the test cross if the original parents had been **AAbb** and **aaBB**?

10. Three different genes were found to affect growth of mice tails. When either a fused tail allele (**F**) was present at one of these loci or a kinky tail allele (**K**) was present at a second, the resulting tails had a kinky appearance. Brachyury (**B**) mice have short tails. Crosses between individuals expressing the different alleles showed that these three genes were linked, close to the histocompatibility H2 gene (**H**). The table shows the distances calculated as the result of a series of pairwise test crosses. Make a map of these four genes showing their relative positions to each other.

Genes involved in test cross	Distance between genes (cM)
F/K	1
F/H	4
B/H	12
F/B	8
K/H	3

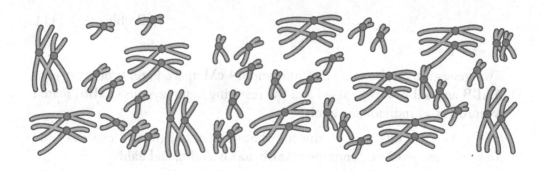

Variation in Chromosomal Number and Structure

The genetic information of a diploid organism is delicately balanced in both content and location. Different genes are found at specific sites on specific chromosomes. Two copies are generally present at corresponding sites on the two homologs. The various alleles of a gene correlate with different phenotypes and some are less efficient than others, occasionally even lethal. It is hardly surprising, therefore, that major changes to a chromosome (e.g. rearranging the genetic layout of a region), or even changing the number of chromosomes, can have a dramatic impact on an organism's phenotype, to the extent of preventing its normal development. Any change to the number or arrangement of chromosomes is known as a **chromosome mutation** or **chromosome aberration**.

This chapter considers chromosome aberrations from two perspectives:

- How they arise

- Their genetic and phenotypic consequences

8.1   Changes in chromosome number: terminology

When an organism or cell has one or more complete sets of chromosomes it is said to be **euploid**. Thus, eukaryotic organisms such as mice and humans, which are normally diploid, can also be referred to as euploid. Likewise, any **polyploid** species with multiple chromosome sets is euploid. Mutations can occur that reduce or increase the number of chromosomes of a set. Any organism or cell with a chromosome number that is *not* an exact multiple of the haploid number of chromosomes is described as **aneuploid** (Figure 8.1).

8.2   Aneuploidy

The most frequent examples of aneuploidy are cases when a single chromosome is either lost from or added to a normal diploid set (i.e. **monosomy** or **trisomy**). The extra or lacking chromosome can be an autosome or a sex chromosome.

Aneuploidy usually results from **non-disjunction** during meiosis, when either a pair of homologous chromosomes fails to segregate during anaphase I or a

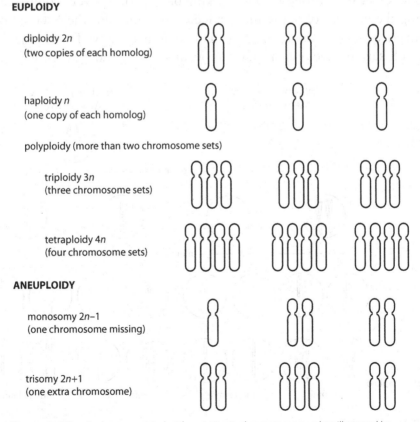

Figure 8.1 Terminology associated with variation in chromosome number, illustrated by considering three homologous chromosome pairs.

pair of chromatids does not separate during anaphase II. Figure 8.2 illustrates this process. Normally a gamete should have one copy of every chromosome. If non-disjunction occurs during the formation of a gamete it becomes **unbalanced** – lacking or containing two copies of a particular chromosome. Following fertilization of such gametes by a normal haploid gamete, monosomic and trisomic zygotes result (Figure 8.2).

Non-disjunction can also occur during mitotic division if sister chromatids fail to separate during anaphase. This results in somatic clones of monosomic or trisomic cells. If mitotic non-disjunction occurs during early embryonic development it produces a **mosaic** individual who has patches of cells with different chromosome numbers.

8.3 Monosomy

The phenotypic consequences to the individual who develops from an aneuploid zygote varies depending upon the chromosome involved, but may be severe. Throughout the animal and plant kingdoms the consequences of either having an extra sex chromosome, or lacking one, are less severe compared with changing the number of autosomes. Indeed monosomy for the X chromosome is relatively common. About 1 in 3000 female humans show **Turner syndrome** (**XO**). Full sexual development fails to occur in these individuals, who have a

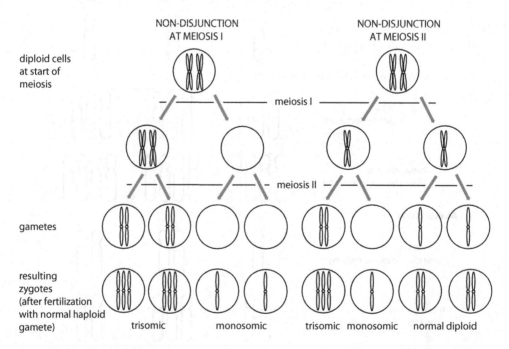

Figure 8.2 Gametes and zygotes produced as the result of non-disjunction during meiosis I or meiosis II. One pair of homologous chromosomes shown.

single X chromosome; ovaries are rudimentary, although external genitalia and internal ducts are present. There can also be various physical abnormalities and health problems, none of which are life threatening. By contrast, monosomy for one of the autosomes is generally lethal in animal species. It is better tolerated in the plant kingdom, although monosomic plants usually show reduced viability compared with their normal diploid counterparts.

So why should monosomic individuals fail to survive, and why is there a difference in toleration of monosomy between animals and plants? The normal situation is for individuals, being diploid, to possess two alleles for each gene. There must, therefore, be many heterozygous situations in which a functional dominant allele is masking a potentially lethal recessive one. Monosomic cells have an unpaired chromosome. If a lethal allele is present at just one locus on this chromosome, then early death of the developing organism occurs.

8.4 Trisomy

In contrast to monosomy, the phenotypic consequences of an extra chromosome are not as severe. In plants, trisomic individuals are often fully viable, although their phenotype may be altered when compared with normal diploid individuals. A classic example of trisomy involves the much studied Jimson weed, *Datura stramonium*. The diploid number of this species is 24. Twelve different varieties are available and each is trisomic for a different chromosome. Interestingly, each trisomy sufficiently alters the development of the seed capsule to produce 12 different forms (Figure 8.3).

The viability of trisomic animals varies, and seems to relate to the size of the chromosome represented in triplicate and, therefore, the number of genes involved. Thus, it is rare to find individuals trisomic for the larger chromosomes. An exception again to this rule is the X chromosome. Males with an extra X chromosome are found in many mammalian species. About 1 in 500 human males are **XXY** and show **Klinefelter syndrome**. Their main difference, compared with XY males, relates to ambiguous sexual characters. The presence of the Y chromosome ensures XXY individuals are phenotypically male. Secondary female sexual development is not, however, totally suppressed; so, for example, slight enlargement of the breasts is common. These problems are accentuated in individuals who occasionally possess more X chromosomes (i.e. XXXY or even XXXXY)!

The only human autosomal trisomy in which a significant number of individuals survive longer than a year is trisomy of chromosome 21, also called Down syndrome (Figure 8.4). Given its relatively high incidence (approximately 1 in 700 live births), it is a condition with which most of us are familiar (Box 8.1). Only two other human trisomies allow individuals to survive to term: trisomy 13

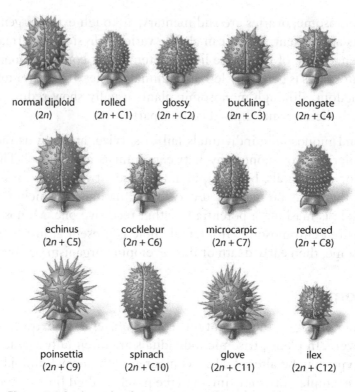

Figure 8.3 Seed capsules of trisomic varieties of Jimson weed ($2n = 24$).

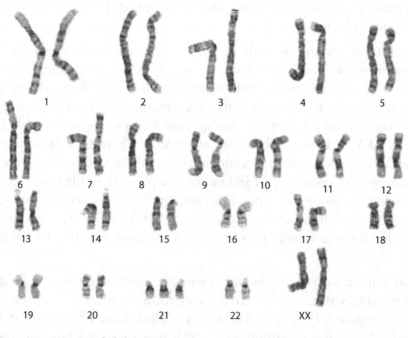

Figure 8.4 Karyotype of a female with Down syndrome (47XX + 21).

BOX 8.1 FEATURES OF DOWN SYNDROME

We may recognize someone as a Down individual as they possess a distinctive phenotype. They are usually of small stature; their eyes are narrow because of a prominent fold of skin in the corner of their eyes; their tongues are large, and they have broad hands with characteristic palm and fingerprints. Indeed, the effect of having an extra chromosome 21 is wide ranging. Physical, psychomotor, and mental development are all retarded. Down people are also prone to respiratory disease and heart malformations, and show a higher incidence of leukemia. Death is frequently due to Alzheimer's disease. This correlates with the locus for one of the genes (*APP*, encoding amyloid precursor protein) that promotes early onset Alzheimer's disease being on chromosome 21.

As with all chromosomal abnormalities, the incidence of Down syndrome increases with maternal age. The reason is unknown, but many scientists would relate the increase that occurs with advancing maternal age to the greater age of the ova. All of a woman's potential eggs are formed by birth and arrested at meiosis I. An older unused egg has had a longer period in which non-disjunction and other defects can occur.

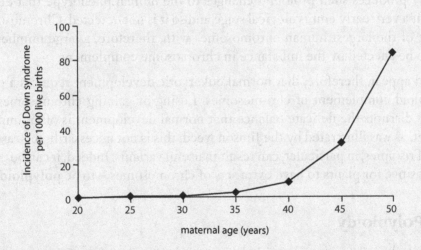

Box 8.1 Figure 1 Incidence of Down syndrome.

All pregnant women in the UK are offered an assessment of their risk of carrying a Down baby. During weeks 11–14 of pregnancy, a nuchal translucency (NT) ultrasound scan measures the thickness of a pocket of fluid at the back of a fetus's neck. This information is combined with maternal age, and a blood test that looks at levels of human chorionic gonadotropin and pregnancy-associated plasma protein A. Further tests are offered if these test results indicate an increased risk (1 in 150 or greater) of the fetus being trisomic for chromosome 21. For example, fetal cells are obtained by amniocentesis or chorionic villi sampling and a karyotype prepared. If three copies of chromosome 21 are present, the prospective parents can decide whether or not to continue with the pregnancy. It is routine for older women to be offered chromosomal karyotyping. Currently, a test to analyze fetal DNA (produced from areas of the placenta) in maternal blood is being developed. This could soon eliminate invasive amniocentesis or chorionic villi sampling, with their attendant miscarriage and occasional fetal damage problems.

Although we can detect the presence of an extra chromosome 21, we cannot predict how severely affected any Down child will be. Many Down people lead happy productive lives, while others are severely mentally and physically handicapped.

(Patau syndrome) and trisomy 18 (Edwards syndrome). Both syndromes result in severe malformations and early death.

Trisomies of all other human chromosomes, except chromosome 1, are known, but the affected fetuses are spontaneously aborted. Trisomy of chromosome 1 probably produces such profound changes to the human phenotype that death occurs at a very early embryological stage and so it is not detected. Chromosome 1 is one of the largest human chromosomes with, therefore, a large number of genes to be affected by the imbalance in chromosome complement.

It would appear, therefore, that normal eukaryotic development requires a precise diploid complement of chromosomes. Losing or gaining chromosomes so severely disrupts the delicate balance that normal development is often impossible. Yet, as was illustrated by the Jimson weed, this is not necessarily the case in plants. Trisomies, in particular, can result in useful variants. Indeed, it can be even more positive for plants to have extra sets of chromosomes – to be **polyploid**.

8.5 Polyploidy

Polyploidy describes a genome with three or more complete sets of chromosomes. It is a widespread phenomenon among plant species: one-third of all genera are believed to contain polyploid species. Polyploid plants tend to grow more vigorously and are larger than their diploid relatives, producing larger flowers and fruit (Figure 8.5). Not too surprisingly, therefore, many commercially grown plants are deliberately produced polyploids! A selection of polyploid plants is listed in Table 8.1. Polyploidy is, however, rare among animals and it is hard to find examples of polyploid animal species. Known polyploid invertebrates include flatworms, earthworms, and brine shrimps. Commercial oysters are triploid, and conveniently do not spawn and so produce unpalatable eggs! Among

(A)   (B)

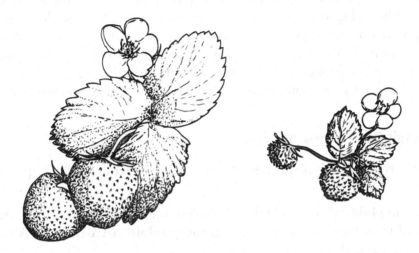

Figure 8.5 A comparison of the cultivated octaploid strawberry (A) and its wild diploid relative (B).

TABLE 8.1 Examples of commercially grown polyploidy plants

PLANT	PLOIDY	PLANT	PLOIDY
Banana	$3n$	Peanut	$4n$
Watermelon	$3n$	Alfalfa	$4n$
Winesap apples	$3n$	Cotton	$4n$
Potato	$3n$	Wheat	$6n$
Tulip	$3n$	Strawberry	$8n$
Coffee	$4n$	Some blackberry varieties	$12n$

vertebrates, a few frogs, toads, salamanders, and lizards are polyploid. In 1999, the first ever polyploid mammal was reported: an Argentinean desert rat, the red viscacha rat, *Tympanoctomys barrerae*, which was found to contain 102, or four sets, of chromosomes in each of its cells.

The higher incidence of polyploid plants compared with animal species is striking. This difference may well relate to the fact that polyploid organisms are often sterile. Plants can generally overcome a fertility problem by reproducing asexually. Few higher animals reproduce asexually. Indeed, any polyploid animal species usually shows **parthenogenesis** (i.e. they produce eggs that develop into new individuals without fertilization). The sterility associated with polyploidy arises from pairing problems during prophase I of meiosis. With multiple chromosome copies, homologs cannot pair properly. Some polyploids have an even number of chromosomes, such as **tetraploids** (4*n*) and **hexaploids** (6*n*), while others have an odd number, such as **triploids** (3*n*). With an even number of chromosomes there is a possibility that homologs might correctly pair. It is impossible for three copies of a homolog to sort themselves into pairs. Segregation of chromosomes cannot, therefore, properly occur at anaphase and so unbalanced gametes are produced with variable numbers of chromosomes (Figure 8.6). These gametes cannot lead to the production of fertile seeds. Hence, the palatability of commercially grown bananas and watermelons – they are triploid and seedless!

8.6   Origins of polyploidy

Two main types of polyploid species can be recognized depending upon their origins:

- **Autopolyploids** possess identical chromosome sets because they are all derived from the same species (the **autotetraploid AAAA**, where **A** represents the haploid chromosome set).

- **Allopolyploids** possess chromosome sets of two or more different types because they are derived from different species (e.g. the **allotetraploid AABB**, where **A** and **B** represent chromosome sets from two different species).

8.7   Producing an autopolyploid

Autopolyploids may arise in several ways:

1. If segregation of all chromosomes fails to occur during meiosis then a diploid gamete is produced. If this is fertilized by a normal haploid gamete, a triploid

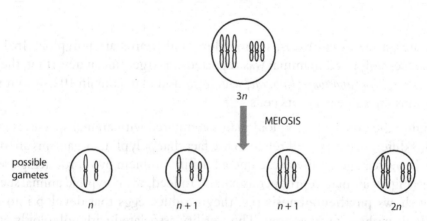

Figure 8.6 Meiotic segregation of chromosomes into gametes in an autotriploid. Two sets of chromosomes are shown. Note that only one in four gametes is likely to have the normal haploid complement.

zygote results. Non-disjunction of the meiotic chromosomes may occur naturally or be induced by heat or cold shock during meiosis or by adding the autumn crocus extract, **colchicine**, to the dividing cells. Colchicine inhibits spindle formation. With no spindle apparatus, homologous pairs of chromosomes cannot separate during anaphase.

2. Two sperm, or two pollen nuclei, can fertilize one ovum; occasionally a triploid human is produced in this way. Usually a triploid human fetus is spontaneously aborted, but, very rarely, one develops to term. However, all systems are so profoundly affected that the baby cannot live more than a few hours or days.

3. An autotriploid is produced by mating a tetraploid with a diploid individual. The latter produces haploid gametes, while the tetraploid forms diploid gametes. Three sets of chromosomes will therefore exist in the zygote.

8.8 Producing an allopolyploid

Allopolyploidy results from the hybridization of two closely related species and subsequent chromosome duplication. Consider Figure 8.7. Species A with three pairs of chromosome crosses with species B with a different set of three chromosome pairs. The resulting hybrid is diploid and sterile. Viable gametes cannot be produced as the chromosomes from the two different species cannot pair during meiosis. The hybrid's propagation relies solely upon asexual reproduction. If, however, the hybrid undergoes a natural or induced chromosomal doubling, a fertile **allotetraploid** or **amphidiploid** (two complete diploid genomes) is produced. There are now two homologous sets of chromosomes. The chromosomes of each separate set pair during meiosis. Normal segregation occurs at anaphase and balanced viable gametes are produced. Allotetraploids can also be produced by the fusion of two unreduced diploid gametes from two different species.

	species A				species B		
chromosome pairs	a_1a_1	a_2a_2	a_3a_3		b_1b_1	b_2b_2	b_3b_3
gametes	a_1	a_2	a_3		b_1	b_2	b_3
sterile diploid			a_1	a_2	a_3	b_1	b_2	b_3
fertile tetraploid			a_1a_1	b_1b_1	a_2a_2	b_2b_2	a_3a_3	b_3b_3

chromosome doubling

Figure 8.7 The origin of an allotetraploid.

There are some well-documented and commercially important examples of alloploidy. Modern cultivated wheat, *Triticum aestivum*, is an allohexaploid, and is believed to be the result of two natural hybridizations and chromosomal doubling events (Figure 8.8). Cultivated cotton, *Gossypium* (various species), is a natural hybrid between an Old World and a wild American species in which colchicine was used to induce a chromosome doubling. In both cases the polyploid form is considerably more prolific than the original diploid species. The commercially useful "triticale" is a designer hybrid between emmer wheat (*Triticum turgidum*) and rye (*Secale cereale*). The aim was to combine wheat's high-yielding character and high protein content of its grain with the flexible growth requirements and high lysine content of rye grains.

8.9   Changes in chromosome structure

The second group of chromosome aberrations that can result in changed phenotypes are structural changes to individual chromosomes. Portions may be deleted or added, or a rearrangement of genetic material may occur, as illustrated in Figure 8.9.

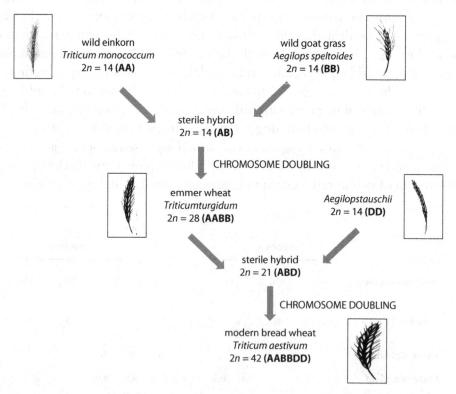

Figure 8.8 The probable evolutionary history of wheat over the last 10,000 years. The letters A, B, and D represent different haploid sets of chromosomes. In many modern varieties the long hairs have been eliminated by selective breeding.

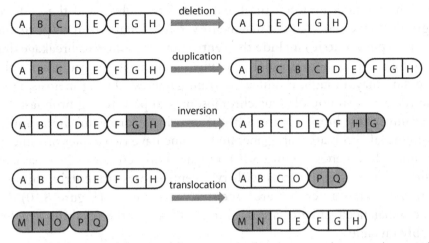

Figure 8.9 Summary of chromosomal rearrangements (letters represent chromosomal segments).

Chromosomal rearrangements occur because of one or more breaks along the axis of a chromosome. Chromosomal breakage can occur spontaneously and is, of course, intrinsic to the process of crossing over during prophase I of meiosis. Breakage rates are increased in cells exposed to certain chemicals or radiation. A broken piece of chromosome readily joins with another broken end. If a breakage and rejoining event does not re-establish the original relationship and, if the alteration occurs during gamete formation, the gametes will contain the structural rearrangement. The changed chromosome will be heritable and may have major effects on an individual's phenotype.

The effect of a chromosomal rearrangement may be immediate or delayed for a generation. The timing of expression is related to the type of structural change. If, for example, genes are missing because of a deletion or a crucial gene's function is abolished because a break point occurred within the gene, then the effect is immediate. If, however, no loss or gain of genetic information occurred then the phenotypic consequences of a rearrangement may not be expressed for one or more generations. These eventual phenotypic changes are a consequence of problems during meiosis in the non-affected individual with the balanced chromosome changes. Many rearrangements cause problems for homolog pairing during prophase I (Section 8.10). These result in gametes either deficient in or containing extra chromosomal material. Individuals resulting from these unbalanced gametes have an increased probability of showing multiple phenotypic changes.

8.10 Changes in the arrangement of genes

An **inversion** occurs when a segment of a chromosome is rotated by 180°. This is believed to occur when a chromosome forms a tight loop and breaks in two places. The segment between the break untwists slightly and the ends rejoin wrongly. The inverted segment may be of variable length. It may (**pericentric**) or may not (**paracentric**) include the centromere. As long as a breakage did not occur within a gene there is usually no effect on phenotype. An individual with an inversion may, though, produce aberrant gametes. During meiosis, in order for the relevant two homologous chromosomes to pair during prophase I, they form an **inversion loop** (Figure 8.10). If crossing over occurs in an inversion loop, abnormal chromatids are generated – some have two copies of some genes while others lack genes (Figure 8.10). Furthermore, crossover in a paracentric inversion generates chromatids with two centromeres (**dicentric** chromatids) and others lacking a centromere (**acentric** chromatids) (Figure 8.10). If any of these unbalanced gametes are used in fertilization, the resulting zygotes are invariably inviable.

Another type of rearrangement happens when breakage occurs simultaneously in two different homologous chromosomes and they exchange segments. This is a **reciprocal translocation** (Figure 8.11). Again, unless breakage abolishes a critical gene function, the presence of a translocation does not directly alter the

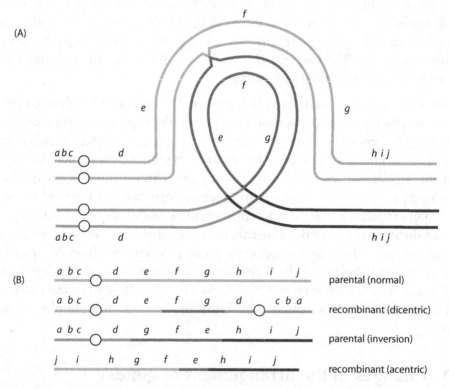

Figure 8.10 Meiosis in an inversion (paracentric) heterozygote (letters represent genes, genes *e/f/g* are inverted in one chromosome). (A) The inversion loop that forms during prophase I when homologous chromosomes pair. (B) Chromatids that result from a crossover in the inversion loop.

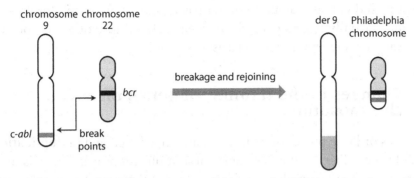

Figure 8.11 Reciprocal translocation between chromosomes 9 and 22 (der = derived chromosome).

viability of an individual. In Figure 8.11 the breaks occurred within genes. Most of chromosome 9 proto-oncogene c-*abl* is transferred to chromosome 22 where a break occurs within gene *bcr*. Disruption of these two genes results in **chronic myelogenous leukemia (CML)**. The small resulting chromosome 22 is referred to as the Philadelphia (Ph) chromosome and is diagnostic of CML.

Chromosomal rearrangements in somatic cells are increasingly being linked to the development of human cancers. It is difficult to know whether the various chromosomal aberrations observed in tumor cells are the cause of the tumor or whether they result from the changed growth activities of the tumor cells. There is, however, growing support for the first hypothesis. In some cancers the associated chromosomal change is always the same. For example, both CML and Burkett's lymphoma have characteristic chromosomal translocations (Table 8.2). In each case the translocation has been shown to activate an **oncogene** – a gene whose product initiates the transformation from a normal differentiated cell to one capable of the uncontrolled growth characteristic of tumor cells.

The presence of a translocation causes pairing problems during meiosis. The result is a proportion of gametes that are unbalanced (i.e. they contain chromosomes with duplicated and deleted segments). When used in fertilization,

TABLE 8.2 Chromosomal changes and human cancers

CANCER	CHROMOSOMAL CHANGE
Chronic myelogenous leukemia	Translocation between chromosomes 9 and 22
Burkitt's lymphoma	Translocation between chromosomes 8 and 14
Ovarian papillary carcinoma	Translocation between chromosomes 6 and 14
Neuroblastoma	Deletion of end of p arm of chromosome 1
Small cell lung carcinoma	Deletion of a section of p arm of chromosome 3
Wilm's tumor	Deletion of end of p arm of chromosome 11

the resulting individual may be partially trisomic or monosomic for a chromosome. Being partially trisomic for chromosome 21, because of a translocation, is another way that a Down individual arises (Box 8.2).

8.11   Changes in the number of genes on a chromosome

Single genes or large pieces of a chromosome may be gained (a **duplication**) or lost (a **deletion**). The commonest cause of a duplication or deletion is unequal crossing over between homologs during meiosis, so that one chromatid gains, and

---

BOX 8.2 PRODUCTION OF FAMILIAL DOWN SYNDROME

Most translocations are reciprocal, with chromosomal segments exchanged. Another kind of translocation involves break points at the extreme ends of the short arms of two non-homologous chromosomes. The small segments are lost, while the longer ones fuse to produce one new large chromosome. This is a **Robertsonian translocation**. It can occur, for example, between chromosomes 14 and 21.

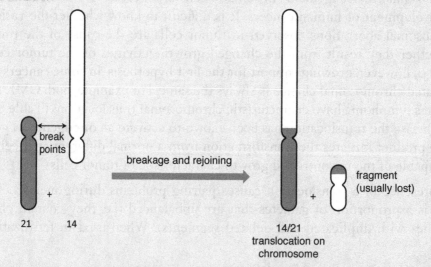

Box 8.2 Figure 1 Robertsonian translocation.

This 14/21 Robertsonian translocation explains the 5% of cases where Down syndrome is inherited or familial. In these cases, one of the parents is a balanced 14/21 translocation. He or she is phenotypically normal, even though there are only 45 chromosomes, as all the relevant genetic information is present. However, at meiosis this individual produces mostly unbalanced gametes (see diagram below). One in six of the gametes will contain two copies of chromosome 21 and when fertilized by a normal haploid gamete result in a Down child.

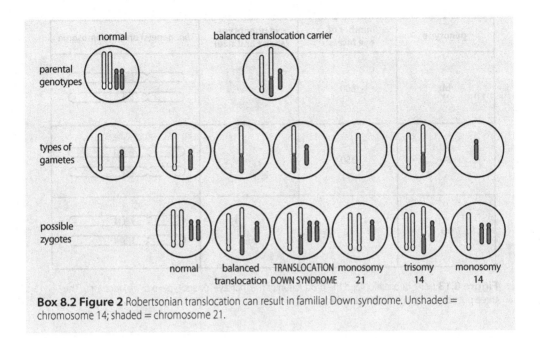

Box 8.2 Figure 2 Robertsonian translocation can result in familial Down syndrome. Unshaded = chromosome 14; shaded = chromosome 21.

the other loses, a region. This can occur when homologous chromosomes mispair (Figure 8.12).

If a gamete carrying a deletion or duplication is viable and fuses with a normal gamete, the individual produced may show a new phenotype; for example, the *Cri du Chat* (French, "cry of the cat") syndrome in humans or the "bar-eye" phenotype of the fruit fly *Drosophila melanogaster*. *Cri du Chat* syndrome is caused by a deletion of much of the short arm of chromosome 5, and is characterized by mental retardation, microcephaly (a small head), and a distinctive cry. Bar-eyed flies have narrow slit-like eyes instead of the normal oval-shaped insect form (Figure 8.13) and duplication of a piece of the X chromosome. The duplication is inherited in a dominant fashion.

It is obviously disadvantageous for the fruit fly to have bar eyes as, with fewer eye facets, its vision is impaired. Gene duplication can, however, play a positive role

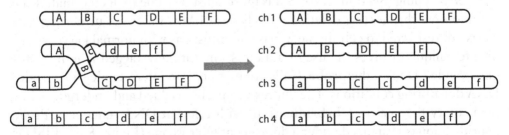

Figure 8.12 The origin of chromosomes with duplicated or deleted regions. Non-sister chromatids are shown mispairing at prophase I of meiosis. Following crossing over, this mispairing produces one chromosome with missing and another with extra genes (ch 2 and 3 in figure).

genotype	number of eye facets	eye phenotype (shape and size)	*bar* gene(s) on X chromosome
bb	800	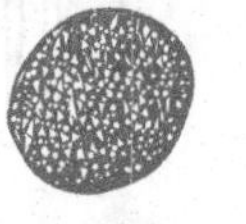	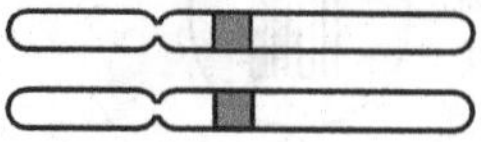
Bb	350		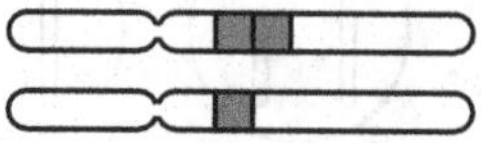
BB	70		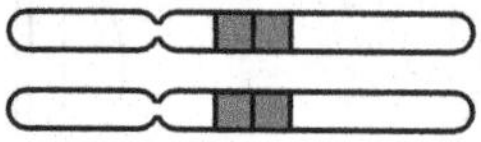

Figure 8.13 Gene duplication and the production of the "bar-eye" phenotype in *D. melanogaster*. The three possible female genotypes are shown.

in cells. Some gene products are needed in large amounts in cells. These genes are present in multiple duplicated copies, such as the gene coding for rRNA that forms an integral part of ribosomes, synthesizing proteins. Gene duplication is believed to have played an important role in evolution. Pairs of genes have been found that have a substantial amount of their DNA sequence in common, such as the protein digestion enzymes trypsin and chymotrypsin, and also the oxygen carriers myoglobin and hemoglobin. Such gene pairs are believed to have arisen from a common ancestor through duplication. The related genes diverged sufficiently during evolution to produce unique products.

8.12   Fluorescent *in situ* hybridization

In recent years, **fluorescent *in situ* hybridization** (**FISH**) has become a common method of examining chromosomes microscopically. Critically, compared with karyotyping (Section 5.1), FISH is more sensitive, able to detect small-scale chromosome changes (e.g. deletions of as little as 1 Mb of DNA). Furthermore, its use is not restricted to cells in metaphase of mitosis, as with normal karyotyping. The technique involves the use of a fluorescent antibody-tagged length of single-stranded DNA – the **probe**. Cells are mounted on microscopic slides under DNA denaturing conditions. The fluorescent probe is added and undergoes complementary base pairing (**hybridization**) with relevant DNA sequences. Probe hybridization is visualized under a fluorescent microscope (Figure 8.14). FISH is used within a range of different contexts; for example, in genetic counseling and prenatal testing when it is useful for identifying missing or extra chromosomes,

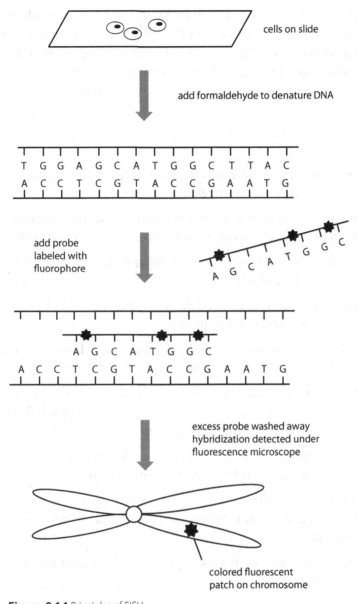

Figure 8.14 Principles of FISH.

deleted, or structural rearrangements of chromosomes. In a research context FISH can, for example, enable study of the spatial and temporal patterns of gene expression as it can detect and localize specific mRNAs within tissue samples.

A disadvantage of FISH is that it depends upon probes that are pre-manufactured. Thus, in chromosome analysis, FISH looks for particular types of aberrations. A new technique of chromosomal analysis enabling full genomic scans – **chromosomal microarrays** – may well replace FISH. During chromosomal microarray analysis, sample DNA (e.g. from a fetus) is hybridized to a microchip containing genomic DNA. The sample and chip DNA are labeled with different fluorescent dyes. The relative intensities of the different colors gives information about deletions, duplications, and other chromosomal abnormalities, and crucially no predictions about possible abnormalities need to be made.

Summary

- If the chromosome complement of a diploid organism is changed it can have profound phenotypic consequences.

- A polyploid individual possesses one or more extra sets of chromosomes, while the cells of an aneuploid individual lack or contain extra single chromosomes.

- Non-disjunction of chromosomes or chromatids during meiosis accounts for the production of aneuploids.

- Monosomic individuals lack one chromosome. In animals this is usually lethal, but it is better tolerated in plants.

- Trisomic individuals possess an extra chromosome. This results in variable viability in animals. In plants trisomy can produce useful variants.

- Polyploidy is a widespread phenomenon in plants. Autopolyploids possess multiple sets of the same chromosome while allopolyploids have chromosome sets from two or more different species.

- Portions of a chromosome may be deleted or duplicated or inverted by 180° or exchanged with segments from another chromosome.

- Structural changes to chromosomes can have a variety of different consequences depending upon the nature of the change and the stage of development when they occur.

- FISH is an increasingly used technique to study numerical and structural changes to chromosomes.

Problems

1. In humans there are 23 chromosomes per haploid set. Aborted fetuses were found to have the following conditions. How many chromosomes would you expect in the somatic cells of each of the following individuals?

   (a) Monosomy 3.

   (b) Triploid.

   (c) Trisomy 16.

(d) Tetrasomy 14.

2. Below is the normal gene sequence on a chromosome:

HIJKL.MNOPQ

What chromosome change resulted in the following chromosomes?

(a) **HIJKNM.LOPQ**

(b) **HIJKL.MNONOPQ**

(c) **HIJKLLK.MNOPQ**

(d) **HIJKL.MNCDEFG**

3. A diploid species with 32 chromosomes is crossed with another with 26 chromosomes.

(a) What would be the number of chromosomes in the hybrid and in the allo-polyploid derived from this hybrid?

(b) Would you expect the hybrid and polyploid offspring to be fertile?

4. Only rarely does an inversion or a reciprocal translocation have a detrimental effect upon the phenotype of an individual. Explain why.

5. What types of gametes, and in what proportions, would an individual trisomic for a chromosome produce?

6. Genes **PQRSTU** are known to be closely linked, but there are doubts as to their correct order. A set of individuals with deletions in this area were obtained. The deletions uncover recessive mutations:

- Deletion 1 uncovers **p**, **q**, and **s**
- Deletion 2 uncovers **p**, **s**, **r**, and **u**
- Deletion 3 uncovers **t** and **u**
- Deletion 4 uncovers **q** and **s**

What is the order of these genes?

7. Two phenotypically normal parents have a child with severe learning difficulties. A karyotype revealed that the child had two normal copies of chromosome 12, but only one normal copy of chromosome 9. The second copy of chromosome 9 had a large section of its short arm replaced by a piece of chromosome 12. Suggest, with reference to chromosomes 9 and 12, the chromosome compositions of the child's parents.

8. When mitosis was examined in root tip cells from an autohexaploid plant species, each cell was found to contain 72 chromosomes. How many chromosomes did each gamete of the original diploid plant contain?

9. An allotetraploid plant has 44 chromosomes in each somatic cell. How many linkage groups are there?

10. A woman with Turner syndrome also shows the X linked condition, red–green color blindness. She had a color blind-father and a mother with normal vision. In which of her parents did non-disjunction of the sex chromosomes occur?

11. How many chromosomes would be found in the somatic cells of an allohexaploid plant derived from three species in which the haploid chromosome number is 9, 11, and 12, respectively?

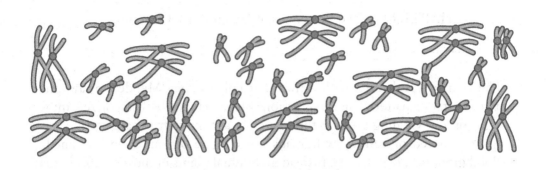

The Principles of Quantitative Genetics

A farmer moved from a fertile plateau at an altitude of 2000 m in the Andes to a new farm, much lower on the Patagonian plains, taking with him a stock of his high-yielding prize potatoes. He confidently looked forward to a good crop at the end of his first year. Instead, he was dismayed by the yield of few and small potatoes. Attributing the low yield to the inevitable, but occasional, bad year, he planted out his seed potatoes again the following year, only to obtain an even poorer harvest. The farmer was bemused. He was aware of the principles of Mendelian genetics and so had expected that the high-yielding gene, or genes, that his potatoes possessed would continue to be passed to subsequent generations and thus ensure him continued high productivity in his new farm. If anything, he had expected that his harvest might have been even better in the lowlands, with their more equitable climate and longer growing season. What he had not fully appreciated was that he had developed his high-yielding variety under a specific set of conditions – high rainfall, a short growing season, and low soil nutrients. His potatoes' high-yielding genotype needed this particular set of environmental conditions for full expression.

The farmer was dealing with a trait that fails to obey Mendelian laws in its inheritance patterns and expression. Plants cannot be categorized into high- or low-yielding types in the same way as tomato fruits are easily recognized as red or yellow, or humans possess one of four clearly defined blood groups: A, B, AB, or O. Instead, individual plants of most species tend to show great variability in their yield; the exact amount lies somewhere on a continuum between two extreme values. Furthermore, plant yield is just one of a wide range of traits of both plants and animals, including humans, which is influenced by a multitude of different factors. Milk production in cows, ear length in corn, and blood pressure in humans are examples of characters that are the consequence of both the genes that individuals possess and a range of different factors in their environment. As the phenotypes of such **complex** or **multifactorial** traits are measured and given a quantitative value, they are also described as **quantitative traits** and the area of genetics that considers their mode of inheritance is referred to as **quantitative genetics**.

This chapter shows how:

- Statistical techniques are used to assess the relative contribution of genetic and environmental factors to the phenotypic expression of quantitative traits.

- Calculated values for the relative input of inheritance and aspects of the environment contribute to the development of plant and animal breeding strategies.

- Many human traits are affected by multiple genetic and environmental factors.

9.1 Multifactorial traits: some definitions

When considering multifactorial traits, individual differences in phenotypes result from:

1. **Genetic factors**, in the form of alternative genotypes among individuals resulting from the expression of two or more genes. Often many genes contribute towards expression of a multifactorial trait, which is recognized in the term **polygenic inheritance**. Thus, it becomes important, but very difficult, to work out how many genes are contributing to the expression of a trait and what are the dominance relationships between alleles at the different loci, and whether gene interactions such as epistasis exist between these alleles.

2. **Environmental factors**, in the form of varying internal and external conditions. Temperature, diet, and parental care are examples of environmental factors that could vary during development of individuals, and so bring about differences in the expression of traits.

For some traits it is differences in genotypes between individuals that have the major influence on phenotype; for other traits, it is variable environmental factors acting on individuals within a population. Geneticists often want to know the relative importance of genotype and environmental factors in contributing to phenotype so that, for example, a pig breeder can know whether he or she is better advised to put efforts into devising an appropriate breeding programme or into developing new feeding strategies in order to increase growth rate in their pigs. To answer such questions geneticists examine patterns of inheritance between generations, as in monohybrid and dihybrid crosses, but the results are presented and analyzed differently. Statistical techniques are used to assess the relative contribution of inheritance and environment to the expression of a trait.

9.2   Representing multifactorial traits

Frequency diagrams or histograms are used to represent the variation shown by multifactorial traits. Typically, such traits are those for which a measurement can be made for each individual in the population under investigation, such as the length of each ear of corn, the weight of each salmon, or the number of eggs laid by a hen each year. Consider an investigation into the variation in tree heights, part of an ecological monitoring programme in Mikumi National Park, Tanzania.

In total, 576 trees were measured. This survey produced a height range from 1.8 to 24.6 m. The height range was divided into 10 equally spaced classes. The number of trees that fell into each class was counted, and this data was plotted with the class sizes on the horizontal axis and the numbers of trees in each class on the vertical axis (Figure. 9.1).

It can be seen that the histogram representing the range of tree heights is bell-shaped; stated more formally, it shows a **normal distribution**. The height of the majority of trees fell within the middle of the range with numbers falling off towards both ends. The frequency diagram for most multifactorial traits approximates to a normal distribution. This important observation means that we can use the properties of a normal distribution to investigate the variation shown by multifactorial traits; we can, for example, come up with an estimation of the relative contribution of genetic and environmental factors to phenotype.

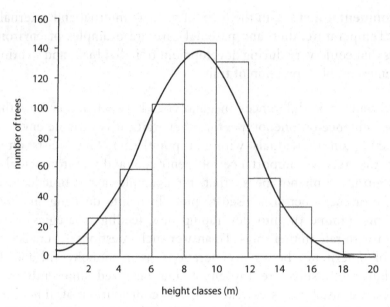

Figure 9.1 Histogram showing the variation in tree height in Mikumi National Park, Tanzania. Each height class includes trees in a 2-m range.

From a sample of measurements, such as tree heights, we can calculate:

- The **mean**, or arithmetic average

- The **variance**, which is an indicator of the spread of the measurements around the mean (i.e. whether the sample shows a small or large range of variation)

- The **standard deviation**, which is another indicator of the range of variation shown by the sample

9.3   Calculating the mean

A normal distribution shows that a set of measurements tends to cluster around a central value. The mean ($\bar{x}$) is a measure of this central value. It is calculated by summing together each individual measurement and dividing the resulting value by the number of measurements:

$$\bar{x} = \frac{\sum x}{n}$$

where $\sum$ indicates "the sum of," x is the individual measurement, $\bar{x}$ is the mean value, and n is the sample size.

The mean measurement of a sample is, by itself, of limited value as it does not tell us anything about the spread of our results or how variable is the expression of a trait. For example, the tusk weights presented in Box 9.1 are very variable.

BOX 9.1 CALCULATING MEAN ELEPHANT TUSK WEIGHT		

The table below shows the calculation of a mean tusk weight found on dead elephants in Mikumi National Park, Tanzania. Each elephant has (generally) a pair of tusks. Thus, the mean weight of the two tusks of a pair was first calculated and then the mean weight of a tusk for all 12 elephants was determined. This was achieved by summing the weights of all the tusks and dividing the summed value by 12 (the sample size) to give an overall mean tusk weight.

Elephant	Weights of individual tusks of a pair (kg)	Mean weight of a tusk in a pair (kg)
1	21.00/21.40	21.20
2	1.20/1.20	1.20
3	1.00/1.00	1.00
4	0.70/0.60	0.65
5	4.00/3.75	3.88
6	7.00/7.50	7.25
7	9.00/9.00	9.00
8	2.10/1.90	2.00
9	1.03/1.02	1.03
10	1.02/1.02	1.02
11	7.20/7.00	7.10
12	1.90/2.10	2.00
		$\Sigma x = 58.33$

$$\bar{x} = \frac{\Sigma x}{n} = \frac{58.33}{12} = 4.86$$

Thus, the mean weight of a tusk found on a dead elephant was 4.86 kg.

Furthermore it is possible for two sets of measurements to share the same mean value but show a great difference in their ranges. Hence, it is important to also have a value for the variance.

9.4   Calculating the variance

The variance indicates the variability within a set of measurements or, stated more formally, the degree to which the measurements diverge from the mean. Thus, the mean is used to produce a value for the variance (σ^2). It is the sum of the squared difference between each measurement and the mean value, divided by one less than the sample size:

$$\sigma^2 = \frac{\Sigma(x - \bar{x})^2}{n - 1}$$

where Σ indicates "the sum of," x is the individual measurements, $\bar{x}$ is the mean value, and n is the sample size.

Box 9.2 shows the calculation of the variance for the set of elephant tusk weights (Box 9.1). This particular variance is a high value, which indicates great variability for this trait between different animals. Estimating values for the variance of different samples has become an important tool in assessing the relative importance of genetic and environmental factors in determining the range of phenotypes of a given trait. Before showing how variance is used in this context, the **standard deviation** will be considered.

9.5   Calculating the standard deviation

The standard deviation (σ) is often preferred to the variance (σ^2) as an indicator of the variability shown by a sample. This preference is because the variance is in squared units, while the standard deviation of a sample is presented in the same

BOX 9.2 CALCULATING THE VARIANCE FOR THE WEIGHTS OF THE ELEPHANT TUSKS

Elephant	Weight of a tusk in a pair in kg (x)	$x - \bar{x}$	$(x - \bar{x})^2$
1	21.20	16.34	267.00
2	1.20	−3.66	13.40
3	1.00	−3.86	14.90
4	0.65	−4.21	17.72
5	3.88	−0.98	0.96
6	7.25	2.39	5.71
7	9.00	4.14	17.14
8	2.00	−2.86	8.18
9	1.03	−3.83	14.67
10	1.02	−3.84	14.75
11	7.10	2.24	5.02
12	2.00	−2.86	8.18
			$\Sigma(x - \bar{x})^2 = 387.63$

$$\sigma^2 = \frac{\Sigma(x - \bar{x})^2}{n - 1} = \frac{387.63}{11} = 35.24$$

Thus, the variance of the weight of a tusk found on a dead elephant is 35.24.

units as the original measurements. The difference between the two is subtle. Variance describes the total spread of a set of measurements around a mean value. The standard deviation indicates what percentage of our measurements will be found within a certain range of the mean. The two terms are, therefore, closely related in concept and also in their method of calculation. The standard deviation is, simply, the square root of the variance.

The calculated value represents **one standard deviation**: 68.26% of all measurements fall within the range of the mean plus or minus one standard deviation; 95.44% of all measurements fall within the mean plus or minus two standard deviations (Figure 9.2). Information about the variability of a sample comes from the specific value for the standard deviation. The larger its value, the greater the range of results. As for variance, this concept needs some figures to illustrate its use! We can calculate the standard deviation for the weight of the Mikumi elephant tusks:

Variance (σ^2) = 35.24 (from Box 9.2)

Standard deviation (σ) = $\sqrt{35.24}$ = 5.94

A value of 5.94 is high for a standard deviation. It is, however, not unexpected given the wide spread of values for a small data set, just 12 elephants. By comparison, the standard deviation for the height of the 576 trees in Mikumi National Park was 3.31.

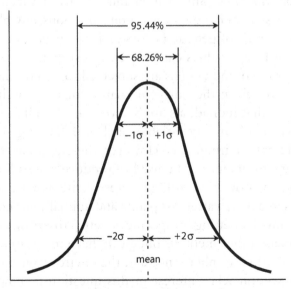

Figure 9.2 Normal distribution curve showing the proportions of the data that are included within one or two standard deviations; 99% of the data lies within the mean plus or minus three standard deviations.

These last few sections have indicated how we do some preliminary analysis of the variation shown by quantitative traits. Once we have values for the mean and variance of a sample we are then in a position to estimate how much of this variation has a genetic (inherited) basis.

9.6 Genes or environment

The variance gives an indication of the range of phenotypic expression for a given trait in the particular population under study. Within quantitative genetics we refer to this as **phenotypic variance** (V_p). The next task is to assess how much of the variation represented by the phenotypic variance might be caused by genetic factors and how much caused by environmental ones. We therefore attempt to break down, or **partition**, the phenotypic variance into two values, a **genetic variance** (V_g) and an **environmental variance** (V_e). Thus:

$$\text{Phenotypic variance} = \text{genetic variance} + \text{environmental variance}$$

$$V_p = V_g + V_e$$

The approach to estimating separate genetic and environmental variances is essentially the same in all plants and non-human animals. If individuals of uniform genotype are nurtured under a range of different environmental conditions, any phenotypic variation should be attributable to environmental factors. The variance calculated from these measurements should, therefore, be an estimate of **environmental variance**. Similarly, if individuals of varying genotype are nurtured under the same set of environmental conditions, any phenotypic variation should be the result of genetic factors and give us a value for the **genetic variance**. Necessarily, any values can only be an estimate, because how can we be certain that a group of individuals are subjected to exactly the same environmental conditions or, given the polygenic nature of most multifactorial traits, how can we be sure that individuals are homozygous for all the relevant genes? There could also be confounding variables such as different genotypes behaving differently in different environments! However, these approaches to controlling genotype or the environment can be used to produce values for the two different variances and so provide useful estimates of the genetic input to various quantitative traits in a wide variety of plants and animals, but not in humans. As these approaches involve setting up specific genetic crosses or carefully controlling the environment, other methods obviously have to be used when assessing the relative contribution of inheritance and the environment in humans. These are discussed in Section 9.11. Interest in this question is, however, high with regard to humans, because it forms part of the hotly debated "nature–nurture" contribution to many of our human traits. A new twist to the "nature–nurture" debate is the recent idea of **epigenetics** – that new environmental experiences that produce novel phenotypes can lead to long term heritable changes. Possible molecular mechanisms (i.e. changes to DNA and chromatin structure) have been suggested for many epigenetic phenomena.

9.7   Partitioning phenotypic variance

A pseudo-Mendelian approach is adopted as the basis for calculating genetic and environmental variance. Individuals from two pure-breeding populations are crossed. A value for the variance in the parental, F_1, and F_2 generations is calculated. From this data it is possible to partition overall, phenotypic, variance into genetic and environmental variances. The principles are best demonstrated by using some data.

Consider some data collected in an early study of quantitative traits. The American plant geneticist Edward East (1879–1938) examined the inheritance of corolla tube length in several strains of the tobacco plant, *Nicotiana longiflora*. Figure 9.3 shows the variation in length of the corolla tube shown by two pure-breeding parental varieties, and by the F_1 and F_2 generations produced by crossing short- and long-flowered plants. A mean and a variance were calculated for each group.

With reference to the data presented in Figure 9.3, we can argue that:

1. The variation in corolla length for each parental group of plants results from variable environmental factors because the two varieties were pure breeding and so should be homozygous for most, if not all, relevant loci. Genetic variance should be, theoretically, zero.

2. The variation among the F_1 offspring should also be solely the result of variable environmental factors because all F_1 individuals should be genetically identical, although this time heterozygous. The expected uniform heterozygosity of the F_1 generation is explained in Figure 9.4.

3. We therefore have three estimates of environmental variance (one for each parental variety and another for the F_1 generation) whose values are averaged. For our corolla tube example this gives us an average value for environmental variance of 8.76.

4. The greater variation in corolla length shown by the F_2 generation results from individuals both experiencing different environmental conditions and possessing different genotypes. When individuals of the F_1 generation are crossed among each other, many different genotypes will segregate out. The variance of 40.96 for the corolla tube example therefore represents both environmental variance (V_e) and genetic variance (V_g).

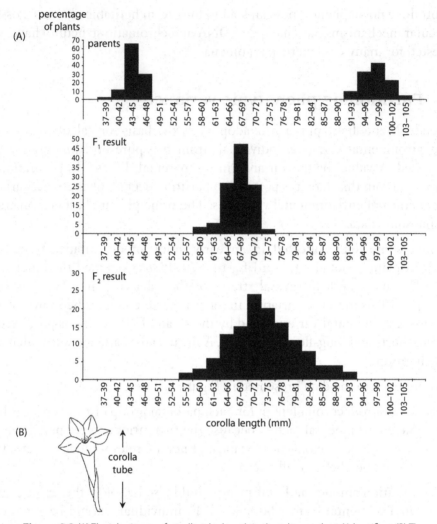

Figure 9.3 (A) The inheritance of corolla tube length in the tobacco plant, *N. longiflora*. (B) The corolla tube of the tobacco plant.

5. Genetic variance (V_g) can therefore be estimated at 32.2. This is calculated using the equation $V_p = V_g + V_e$. We had values for V_p and V_e, and so can rearrange the equation to produce a value for V_g:

$$V_p = V_g + V_e$$

$$40.96 = V_g + 8.76$$

Rearranging the equation:

$$V_g = 40.96 - 8.76 = 32.2$$

The genetic variance (V_g) is therefore much greater than the environmental variance (V_e). Thus, for this F_2 population, we can argue that the observed

genotypes of pure-
breeding parents   AABBCCDD   ×   aabbccdd

gametes   ABCD   ×   abcd

F_1 genotype   AaBbCcDd

Figure 9.4 Explaining the uniform heterozygosity of the F_1 generation in Figure 9.3. It is hypothesized that four genes (**A/a**, **B/b**, **C/c**, and **D/d**) control corolla length.

variation in corolla length is largely the result of genetic differences between individuals or, as the next section explains, approximately 80% of the variability in corolla length among the particular population of tobacco plants being studied is attributable to genetic causes.

9.8 Heritability index

The proportion of the total phenotypic variation attributable to genetic differences between individuals is often formally recognized as the **heritability index** (H^2). We use values gained for genotypic and phenotypic variance to calculate this index. It is equal to the genotypic variance divided by the phenotypic variance:

$$\text{Heritability index } (H^2) = \frac{V_g}{V_p}$$

For the corolla tube example used in Section 9.7:

$$\text{Heritability index } (H^2) = \frac{32.2}{40.96} = 0.79$$

A value of 0.79 for the heritability of our tobacco corolla length example indicates a strong genetic input to the expression of this character in the population under study. Heritability of a trait ranges from 0 to 1. A heritability index of 0 indicates that none of the variation in phenotype among individuals in a population is attributable to genetic causes; a value of 1 suggests that all the observed phenotypic variation has a genetic basis. Table 9.1 shows heritability values for a selection of different traits in a range of species.

At first sight a value for the heritability index would seem an extremely useful thing to know. Consider a dairy farmer, keen to increase the milk yield from his herd. He needs to know how much of the current variability in milk production by his cows is environmentally induced and how much is a result of genetic factors. If the farmer had access to a heritability index value for milk production then he or she would surely know whether a selective breeding programme is

TABLE 9.1 Heritability estimates for a selection of agriculturally important traits (values are generally averaged for a number of populations in differing environmental conditions)

TRAIT	HERITABILITY
Egg weight in poultry	0.55
Egg production (to 72 weeks) in poultry	0.10
Age at first laying	0.52
Litter size in pigs	0.05
Fleece weight in sheep	0.40
Milk production in cattle	0.33
Plant height in maize	0.70
Ear length in maize	0.17
Yield in maize	0.25
Height in spring wheat	0.72
Tiller number in winter wheat	0.36

likely to be successful in raising milk yields, because a high value indicates a strong genetic component and vice versa. Great care, however, has to be exercised when using a heritability index.

Any heritability index is sample and circumstance specific. The variances used to calculate a heritability index are for the particular population under investigation and the conditions prevailing at the time the measurements are made. A heritability index therefore represents how much of the observed variation in phenotype is attributable to genetic factors under the specific set of prevailing conditions. It is impossible to obtain an absolute value of the heritability index for a given trait.

The heritability index we have so far been discussing is often referred to as **broad-sense heritability**. The genetic contribution to the expression of a phenotype encompasses different types of gene actions and interactions (i.e. alleles may be completely or incompletely dominant, or there may be epistatic interactions between genes). Genetic variance can be partitioned into different components which recognize these different types of gene actions.

9.9 Partitioning genetic variance $V_p = V_g + V_e$

Genetic variance (V_g) can be split into three major subcomponents that reflect the different ways in which gene action and interaction can contribute to the phenotype of an individual. The three subcomponents are:

- **Additive genetic variance (V_a)**

- **Dominance variance** (V_d)
- **Interactive variance** (V_i)

The **additive genetic variance** relates to alleles at a locus being incompletely dominant, which often influences phenotypes such as height, color, or mass, in an additive or quantitative fashion. Consider the genetic basis of snapdragon flower color in relation to pigment levels in the petals (discussed in Section 3.2). The three main colors of red, pink, and white are determined by two incompletely dominant alleles at one locus. The additive relationship between genotype and phenotype is summarized in Table 9.2.

If the same **additive** principle applies to the situation where a number of genes contribute to the expression of a particular trait then a wide range of different phenotypes results from the different genotypes (and this is before environmental factors are taken into consideration!). Consider a theoretical situation where two genes, **A** and **B**, control wheat yield (it is actually controlled by many more genes), and where there are two alleles at each locus that contribute to yield in the following relative manner:

- **A** allele = 6 units
- **a** allele = 3 units
- **B** allele = 2 units
- **b** allele = 1 unit

Table 9.3 shows the phenotypic distribution in the F_2 generation when pure-breeding **AABB** and **aabb** parents are crossed, and the resulting F_1 heterozygotes interbred. It can be seen that *just* with two genes, and two alleles at each locus, there are nine different genotypes and so phenotypes. The range of possible phenotypes is greatly expanded if there are more genes contributing to expression of a trait and especially so if at least some of these genes possess multiple alleles.

The **dominant genetic variance** represents the variability in phenotypic expression of a trait that is possible by the action of those contributing loci whose alleles show a definite dominant/recessive relationship. Under these circumstances, the

TABLE 9.2 Relationship between genotype and snapdragon petal colors in terms of numbers of pigment units (allele symbols: **W** = white petals and **R** = red petals)

GENOTYPE	PIGMENT UNITS	PETAL COLOR
WW	None	White
WR	One	Pink
RR	Two	Red

TABLE 9.3 The effect of additive alleles on phenotype: the additive influence of two genes and their alleles on wheat yield (contribution by alleles: **A** = 6 units, **a** = 3 units, **B** = 2 units, **b** = 1 unit; F_1 heterozygotes, obtained from crossing **AABB** and **aabb** parents, were interbred)

GENOTYPE	PROPORTIONS IN F_2	PHENOTYPE (YIELD UNITS)
AABB	1	16
AABb	2	15
AaBb	4	12
AaBB	2	13
Aabb	2	11
AAbb	1	14
aaBB	2	10
aaBb	1	9
aabb	1	8

heterozygous genotype has the same influence on final phenotypic expression as a homozygous dominant genotype. While there may be considerable genotypic variation in a population, the corresponding phenotypic variation is less (i.e. homozygous and heterozygous individuals produce the same phenotype).

Finally, epistatic interactions may occur among alleles at different loci contributing to expression of a given trait (discussed in Section 4.5). The existence of epistasis adds another source of variability when considering the influence of genetic differences between individuals upon phenotypic expression of a trait and is represented by the **interactive variance**.

In summary, therefore, the genetic variance (V_g) is partitioned as:

$$V_g = V_a + V_d + V_i$$

The total, or phenotypic, variance (V_p) is partitioned thus:

$$V_p = V_e + V_a + V_d + V_i$$

9.10   Narrow-sense heritability

Of the three genetic variances discussed in the previous section, additive genetic variance is the most useful for a plant or animal breeder. Many commercially important traits are influenced by genes acting in an additive way. Thus, for the farmer wanting to know whether a breeding programme is likely to increase the amount of fleece his sheep produce or reduce the fat content of his meat, it is crucial to know the additive genetic variance (V_a). This enables a **narrow-sense heritability** (h^2) to be calculated:

$$\text{Narrow-sense heritability } (h^2) = \frac{V_a}{V_p}$$

As with the broad-sense heritability, a high value indicates that much of the variation in observed phenotype results from differences in genotype at additive loci among individuals in the sample population. In contrast, a low value for the heritability index suggests that environmental factors have a greater influence on expression of the trait under consideration.

It is relatively easy to obtain values for phenotypic, environmental and total genetic variances. It is, however, more difficult to design experiments to estimate additive genetic variance and thus be able to calculate a value for narrow-sense heritability. One useful experimental strategy for obtaining an estimate of the narrow-sense heritability will be outlined. It involves performing a one-genera-tion selective breeding programme. If genetic variability for a trait exists among individuals in a population, then it should be possible to change the spectrum of phenotypic expression of that trait by restricting breeding to those individuals which show the desired phenotypes. The extent to which the phenotype changes in one generation is known as the **selection response** (*R*). This can be used to produce an estimate of the narrow-sense heritability, as explained in Box 9.3, and thus an indication as to whether a selective breeding programme is likely to modify the phenotypic expression of a trait in the desired direction.

9.11 Investigating multifactorial traits in humans

It has become increasingly obvious over the last couple of decades that many common human diseases, such as cancer, heart disease, diabetes, schizophrenia, and Alzheimer's disease, are the result of a complex interplay between genetic and environmental factors. Thus, there is much current interest in analyzing quantitative traits in humans. For various reasons, however, the approach is different compared with studying such characters in non-human animals and plants. First, as suggested in Section 9.6, traditional heritability studies, in which controlled matings are set up, are not possible in humans. Instead, human twins have proved informative subjects for assessing the relative influence of genes and the environment in the formation of a range of human traits. Second, the mode of expression of many human multifactorial traits is different.

Hitherto, the traits we have considered in a range of animals and plants have shown a gradation of expression from, for example, low to high yield in agricul-ture. A variety of human traits, such as height or hypertension, do show a similar gradation (Figure 9.5). Many disease conditions are, however, different. They represent an either/or situation – an individual either suffers from or is free of a disease. There is, instead, a gradation of susceptibility or **liability** to developing a

BOX 9.3 CALCULATING NARROW-SENSE HERITABILITY

Consider the upper graph which shows the distribution of mass among a sample of 100 salmon on a fish farm. These salmon can be regarded as the parental generation.

Aim: To increase x_P (mean mass of salmon)

Method: Select and interbreed individuals that fall in the shaded section: x_S is the mean mass of the selected fish

Result: Shown in the lower graph: x_N is the mean mass of the offspring

- Selection differential $(S) = x_S - x_P$

- Selection response $(R) = x_N - x_P$

- Narrow-sense heritability $= \dfrac{R}{S}$

The larger the heritability value, the more likely that a selective breeding programme will be successful.

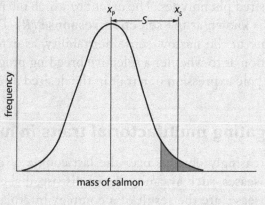

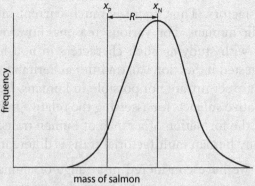

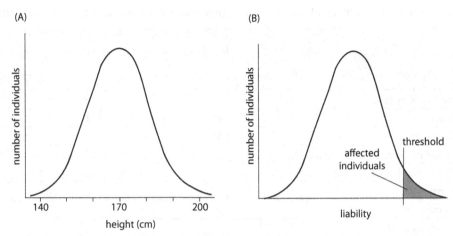

Figure 9.5 Different patterns of expression of human multifactorial trait. (A) distribution of height in the English population. (B) General population liability to express cleft lip and palate.

disease such as diabetes. Once a **threshold** is reached, the condition is expressed (Figure 9.5). The underlying quantitative principle, however, remains the same, in the sense that a number of genetic and environmental factors influence a tendency to express the disease, or some other trait. Any individual who both inherits a particular combination of alleles at various loci and experiences one or more unfavorable environment factors, expresses the phenotype.

When using **twin studies** to measure the relative input of genetic and environmental factors in determining expression of a wide range of human quantitative traits comparisons are made of the frequency of expression of different traits in **monozygous** (or identical) and **dizygous** (or fraternal) twins. Monozygous twins originate when a developing embryo splits to form two separate ones. The resulting individuals are genetically identical, examples of natural clones. Dizygous twins are the result of two separate fertilization events. Thus, as with any two siblings, dizygous twins share just half their genes. As any pair of twins share broadly the same environment, we are therefore in a position to assess the genetic contribution made to a characteristic. To achieve this, comparisons are made of the frequencies of expression of different traits in monozygous and dizygous twins, and **concordance** values are calculated. Twins are said to be:

- **Concordant** for a given trait if both or neither of them express it
- **Discordant** if only one shows the trait and the other does not

For traits totally determined by genotype, concordance rates should be 100% for monozygous twins and 50% for dizygous twins. Values will decrease the greater the environmental input, but always be higher for monozygous twins if genetic

TABLE 9.4 Concordance rates in monozygous and dizygous twins for selected traits

TRAIT	CONCORDANCE RATE (%)	
	MONOZYGOUS TWINS	DIZYGOUS TWINS
Alzheimer's disease	58	26
Cancer	17	11
Cleft palate	35	5
Diabetes mellitus (insulin dependent)	35	6
Diabetes mellitus (non-insulin dependent)	100	10
Epilepsy	37	10
Hypertension	30	10
Ischemic heart disease	19	8
Leprosy	60	20
Manic depression	70	15
Measles	76	56
Multiple sclerosis	25	6
Pyloric stenosis	15	2
Schizophrenia	45	12
Spina bifida	6	3
Tuberculosis	87	26

factors are involved (Table 9.4). If similarly high concordance values are obtained for both monozygous and dizygous twins, this indicates a strong environmental input to the expression of condition, as in the high concordance rates for the development of measles (Table 9.4). The shared environment of both kinds of twins exposes them equally to the infectious agent. As an extension to conclusions drawn from twin studies, any environmental effects can be checked out in the few studies that have been possible on twins reared apart. Do concordance rates remain high when environments are different?

There are, however, limits to the information that can be extracted from twin studies. They can never do more than indicate the extent to which a multifactorial condition *may* be caused by genetic factors. If we are to make inroads into understanding and treating some of the major diseases that afflict us, specific genes need to be identified. The application of recent technological advances in molecular biology to the mapping of genes involved in the expression of multifactorial traits, or **quantitative trait loci (QTLs)**, are yielding exciting results. A **genome-wide association study (GWAS)** is conducted. For a given trait the DNA from thousands of individuals expressing and a matched sample of thousands of non-expressing individuals is examined for DNA sequence variants that are associated with the trait. A GWAS typically focuses on searching for associations between **single nucleotide polymorphisms (SNPs)** and a major disease trait. These SNPs or single nucleotide differences between expressing and non-expressing individuals highlight an area of the genome where a gene influencing the trait is likely to be found and thus initiates intense analysis of surrounding DNA for candidate genes. The first successful GWAS was published in 2005 and identified a gene involved in age-related macula degeneration (leading to late-onset loss of vision). Interestingly, the identified gene produces a protein involved in regulating inflammation. Few previously thought that inflammation might contribute so significantly to this type of blindness. This new era of identifying QTLs produces the real hope that treatment and eventually cure of some of mankind's major killer diseases may become possible. A DNA test early in life might perhaps identify individuals with high susceptibilities to certain diseases and thus allow the possibility of preventive treatments. Gene therapy might even provide the ultimate cure.

Summary

- The expression of many traits in living organisms depends upon both genetic and environmental factors.

- The relative contribution of genetic and environmental factors to the expression of multifactorial traits is assessed by the use of statistical techniques.

- Measurements are made of the variation shown by a multifactorial trait; a mean, variance, and standard deviation are then calculated.

- The phenotypic variance represents the range of phenotypic expression for a given trait.

- The genetic and environmental variances indicate how much of the variability in expression between individuals of a sample is the result of genetic and environmental factors, respectively.

- The broad-sense heritability index represents the proportion of the total phenotypic variation attributable to genetic factors. A high value suggests that a selective breeding programme would be effective in achieving a desired phenotype within a group of individuals. The narrow-sense heritability index is sometimes more informative for this purpose.

- An assessment of the relative input of genetic and environmental factors in determining traits in non-human animals and plants is gained from the results of experiments controlling matings and environmental conditions. In humans such information arises from twin studies.

- Geneticists are beginning to identify genes (QTLs) involved in human multifactorial traits.

Problems

1. This question considers a normal distribution.

   (a) What term is given to the value along the x-axis that corresponds to the peak of the distribution?

   (b) If two normal distributions have the same mean values, but different variances, which distribution is the broader?

   (c) What proportions of the population are expected to lie within one, and within two, standard deviations of the mean?

2. The table below represents mean blue tit nestling weight for 10 nests in a mature oak wood and for 10 nests in a marginal scrubland habitat:

Nest	Mean nestling weights in woodland habitat (g)	Mean nestling weights in scrubland habitat (g)
1	10.4	9.01
2	10.2	9.86
3	10.6	9.71
4	9.4	8.84
5	11.0	9.33
6	10.8	10.16
7	11.3	10.8
8	10.3	8.94
9	10.3	9.66
10	10.8	9.43

   Calculate:

   (a) The overall mean weight of the nestlings from the two populations.

   (b) The variance in weight of the nestlings from the two populations.

   (c) Which population has the larger standard deviation?

3. When studying the heritability of a quantitative trait, we commonly analyze its expression in the F_1 and F_2 generations obtained after crossing two highly inbred strains. Which set of progeny provides data for an estimation of the genetic variance?

4. A project was launched to investigate the number of leaf blades produced by different varieties of maize prior to the flower spike. In a cross between two commonly cultivated inbred varieties, the resulting plants showed a variance in leaf number prior to spike production of 1.72 in the F_1 generation and of 4.96 in the F_2 generation.

   (a) What is the broad-sense heritability in leaf number?

(b) If a farmer wanted to breed a new variety that produced less leaf blades before flowering, does this heritability value suggest a selective breeding programme would be effective?

5. Two different varieties of mangoes produce fruit with the same mean weight of 430 g. One variety shows a low variance for fruit weight, while the other possesses a much higher variance.

(a) What are the reasons for the differences in variance?

(b) If you were a commercial mango grower, which variety would you grow and why?

(c) If you wanted to develop a variety that produced heavier fruit, would you choose the variety with the low or high variance? Explain your choice.

6. Values of the time it took 500 schoolchildren to find their way to the centre of a maze showed an approximately normal distribution, with a mean of 10 min and a standard deviation of 3.2 min. What proportion of the group took:

(a) Longer than 16.4 min?

(b) Above 6.8 min?

(c) Less than 6.8 min?

7. Selective breeding is a common strategy for altering quantitative traits. Suppose a plant breeder started with a wild population of blueberries and subjected the plants to many generations of selective breeding, with the goal of obtaining plants bearing larger berries. How would you expect the heritability of this trait in his cultivated populations to change from that in the original wild population after many generations of selective breeding?

8. Assume three key genes determine the weight of pumpkin fruit: A/a, B/b, and C/c. A pumpkin of weight 3 kg and genotype **aabbcc** was crossed with a pumpkin of 6 kg and genotype **AABBCC**. If the presence of each dominant allele results in the addition of an extra 500 g to the developing fruit compared with the recessive homozygote, what weights of pumpkins would you expect in the:

(a) F_1 generation?

(b) F_2 generation?

9. The following variances were calculated for leaf width in a population of cowslips beside a busy motorway: additive genetic variance = 4.2; dominant variance = 1.6; interactive variance = 0.3; environmental variance = 2.7. Calculate the:

(a) Broad-sense heritability of this trait.

(b) Narrow-sense heritability of this trait.

10. A salmon breeder wants to increase the rate of growth of his stock. He therefore chooses to breed from the fish achieving the greatest length by 8 weeks. The mean length was 13 cm and so he decided to breed from individuals that had reached 18 cm or more by 8 weeks. If the mean length of salmon by 8 weeks in the next generation was 15.5 cm:

(a) Estimate the narrow-sense heritability for this trait.

(b) Advise the breeder of the feasibility of his plan to increase the size of his salmon by selective breeding.

11. A farmer has recently acquired a herd of boar and wishes to increase the average body weight. Her current herd has a mean body weight of 325 kg. The farmer decides to breed only from her six heaviest animals who have a mean weight of 350 kg. They produce 39 offspring with a mean weight of 333 kg. Does this indicate a heritable component to body mass among the herd and, so, would you advise the farmer to continue selective breeding to achieve her goal of increasing the average body weight of the herd?

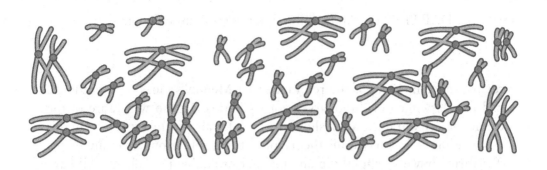

An Introduction to Population Genetics

The focus of population genetics is obviously upon populations. This contrasts with the main concerns of Mendelian transmission genetics, described in earlier chapters, and, to some extent, the quantitative genetics of Chapter 9, where the focus is upon the genotype of individuals and the distribution of genotypes resulting from a single mating. Population geneticists investigate the patterns of genetic variation shown by individuals within different groups, or populations. They are interested in questions such as:

- How much genetic variation exists in a given population?

- What processes control the amount of observed variation in a population?

- How might this genetic variation change with time?

- What processes are responsible for producing genetic divergence between populations?

Answers to such questions have implications for both evolutionary theory and conservation policy. Indeed, the impetus for the development of the discipline of population genetics was the realization, soon after their reemergence in 1900, of the relevance that Mendel's ideas on inheritance had for Darwinian evolutionary theory. Charles Darwin was aware of the tremendous natural variation shown by individuals within a species. On this observation he had based his theory of **natural selection**: "that any being, if it varied however slightly in any manner profitable in itself ... will have a better chance of surviving ... [and] success in leaving progeny." Darwin was, however, unaware of any underlying mechanism that could account for the morphological (phenotypic) variation he observed. The rediscovery of Mendel's work, 19 years after Darwin's death, and the subsequent development of the concept of the gene and alleles, established an understanding of how the mechanism of inheritance could lead to variation. During the early 1900s, the British mathematician Godfrey Hardy (1877–1947) and the German physician Wilhelm Weinberg (1862–1937) each independently produced a set of equations enabling the variation present in populations to be expressed in terms of allele and genotype frequencies.

This chapter:

- Shows how the variation present in a population can be analyzed in terms of allele and genotype frequencies

- Discusses the impact upon the observed variation of crucial features of the population such as its size, mating patterns, or selective forces

10.1 Population genetics: some definitions

A **population** can be defined as a local group of a single species within which mating is actually or potentially occurring. The interbreeding individuals share a common set of genes referred to as the **gene pool**. It is the characteristics of the gene pool that are studied in population genetics (e.g. **frequencies** of different alleles and resulting genotypes in the sample group). Remember that a frequency refers to a proportion, and always ranges in value between 0 and 1. For example, if 19% of people in a group are left-handed, then the frequency of left-handedness in this group is 0.19. The first step in studying the variation present in a population is to measure the frequency of different alleles for individual genes. Population geneticists are often interested in knowing how many loci are **polymorphic** (i.e. possess two or more different alleles with the frequency of the rarest exceeding 0.1%). The more loci that are polymorphic, the more variation the population shows.

10.2   Calculating allele frequencies

Allele frequencies may be calculated in one of two ways. The simplest method is when each of the possible genotypes at a locus produces a separate phenotype. In this case, allele frequencies can be calculated directly from the observed numbers of different phenotypes. Such is the case with the autosomally inherited MN blood groups in humans, described below. The second method uses the recessive genotype frequency as a starting point to deduce allele frequencies, as shown in subsequent sections.

Human blood can be classified in many different ways. The M and N blood groups relate to the M and N antigens produced by a gene on chromosome 4 (Table 10.1). Note that these two blood group alleles are co-dominant so that each possible genotype produces a different phenotype. The two alleles are generally referred to more simply as **M** and **N**.

As each genotype produces a different phenotype we can determine the frequencies of the two alleles directly from collected data. Box 10.1 shows the distribution of MN blood groups in a sample of 1000 randomly selected individuals, and how the frequencies of the **M** and **N** alleles can be calculated from this data.

The frequency of the different MN blood groups varies enormously among different human populations (Table 10.2). This illustrates how variable in their allele frequencies geographically distinct populations can be. As there seems to be no strong selective force acting on blood group antigens, the most likely reason for the range of allele frequencies shown by the different populations is chance fluctuation (see Section 10.9). Another plausible explanation could be that the **M** allele confers an advantage on individuals in a cold climate, while it is more favorable to possess **N** alleles in a hot climate!

10.3   The Hardy–Weinberg law

In the case of the MN blood groups, because the **M** and **N** alleles are co-dominant, each genotype produces a distinct phenotype. However, if one allele is recessive, then the heterozygote possesses an identical phenotype to the homozygous

TABLE 10.1 Relationship between MN blood group genotypes and phenotypes (blood group and antigens within red blood cell membranes).

GENOTYPE	BLOOD GROUP	ANTIGEN (GLYCOPROTEIN) PRESENT IN MEMBRANES OF RED BLOOD CELLS
MM	M	M antigen
MN	MN	M and N antigens
NN	N	N antigen

BOX 10.1 CALCULATING ALLELE FREQUENCIES

$$\text{Frequency of each allele} = \frac{\text{number of copies of an allele}}{\text{total number of alleles}}$$

Phenotype	Genotype	Number of individuals	Number of alleles	
			M	N
M	MM	360	360 × 2	
MN	MN	480	480	480
N	NN	160		160 × 2

Total number of alleles = 2000 (each of the 1000 individuals possess two MN blood group alleles)

$$\text{Frequency of } M = \frac{720 + 480}{2000} = \frac{1200}{2000} = 0.6$$

$$\text{Frequency of } N = \frac{480 + 320}{2000} = \frac{800}{2000} = 0.4$$

TABLE 10.2 Frequencies of **M** and **N** alleles in different populations

POPULATION	M ALLELE FREQUENCY	N ALLELE FREQUENCY
Australian aborigines	0.178	0.822
Japanese Ainus	1.430	0.570
US blacks	0.532	0.468
US whites	0.540	0.460
US Indians	0.776	0.224
Greenland Eskimos	0.913	0.087

dominant individual. In such cases, it is impossible to directly determine allele frequencies. Godfrey Hardy and Wilhelm Weinberg each independently developed a mathematical model that enables us to calculate allele frequencies in such cases. Their work is recognized in the **Hardy–Weinberg law**, which is one of the fundamental concepts in population genetics – as important in its context as Mendelian ratios in transmission genetics. The Hardy–Weinberg law gives us a way of determining:

- Genotype frequencies if allele frequencies are known

- Allele frequencies if certain genotype frequencies are known

According to the Hardy–Weinberg law, the three possible genotypes produced by one gene with two alleles will be present in a population in the following proportions:

$$p^2 + 2pq + q^2$$

where p is the frequency of the dominant allele, q is the frequency of the recessive allele, p^2 is the frequency of the homozygous dominant genotype, $2pq$ is the frequency of the heterozygote genotype frequency, and q^2 is the frequency of the homozygous recessive genotype.

Furthermore, if certain conditions exist within a population the genotype frequencies add up to unity:

$$p^2 + 2pq + q^2 = 1$$

These conditions are that:

- The population under consideration is large (or large enough) that sampling error is negligible.

- No genotype possesses a selective advantage or disadvantage (i.e. that all genotypes are equally viable and fertile).

- There is an absence of other factors, such as mutation, migration, or genetic drift, which could favor certain genotypes.

If all these conditions hold, then the population is said to be in **genetic** or **Hardy–Weinberg equilibrium**, and frequencies of different alleles and genotypes will not change from one generation to the next. In developing their useful equation for calculating genotype frequencies within a population, Hardy and Weinberg both applied basic rules of Mendelian inheritance. A summary of the derivation of the Hardy–Weinberg equation is shown in Box 10.2.

BOX 10.2 DERIVATION OF THE HARDY–WEINBERG EQUATION

Consider a gene with two alleles, **A** and **a**. In population genetics p represents the frequency of the dominant allele (**A**) and q represents the frequency of the recessive allele (**a**). Starting with this premise, the Mendelian principles of inheritance that result from the processes of meiosis and sexual reproduction enabled Hardy and Weinberg each independently to deduce that genotype frequencies in any one generation can be expressed as $p^2 + 2pq + q^2 = 1$. A Punnett square best shows the derivation of this relationship:

Gametes	p (A)	p (a)
p (A)	p^2 (**AA**)	pq (**Aa**)
q (a)	pq (**Aa**)	q^2 (**aa**)

Therefore, with allele frequencies of p and q among the gametes of each parent, the expected genotype frequencies in the next generation will be $p^2 + 2pq + q^2$.

10.4   Calculating genotype frequencies

If we know the allele frequencies in a population, then the frequencies of the resulting genotypes can easily be determined. Consider a population of rats and the gene for resistance to the poison warfarin. If 70% of the alleles at this locus are the dominant resistant allele (**R**) and the remaining 30% the recessive non-resistant one (**r**), then:

$p = 0.7$

$q = 0.3$

We can now use the Hardy–Weinberg equation, $p^2 + 2pq + q^2 = 1$, to calculate the distribution of the genotypes expected in this and any subsequent generations, assuming equilibrium conditions exist:

- Frequency of the dominant homozygote $= p^2 = (0.7)^2 = 0.49$
- Frequency of the heterozygote $= 2pq = 2 \times 0.7 \times 0.3 = 0.42$
- Frequency of the recessive homozygote $= q^2 = (0.3)^2 = 0.09$

Equilibrium is confirmed because these calculated frequencies add up to unity:

$0.49 + 0.42 + 0.09 = 1.00$

We rarely, however, initially know allele frequencies in a population. A more realistic situation is to have phenotype data (i.e. to know how many individuals possess the dominant phenotype and how many the recessive phenotype). It is possible, though, to determine allele frequencies from such data and to then use these to predict genotype frequencies. The starting point in these calculations is to use the frequency of individuals showing a recessive phenotype.

Consider again the Hardy–Weinberg equation and corresponding phenotypes:

$$p^2 + 2pq \quad + \quad q^2 \qquad = 1$$

dominant   recessive
phenotypes   phenotype

To calculate genotype frequencies from phenotype data, the following procedure can be followed:

- First calculate the frequency of the recessive allele (q) from the recessive phenotype (and also genotype) frequency, q^2:

$q = \sqrt{q^2}$

- Having obtained a value for the recessive allele, it is possible to determine the dominant allele frequency, because:

$p + q = 1$

$p = 1 - q$

- Once values for both the dominant and recessive alleles have been calculated, the Hardy–Weinberg equation can be used to deduce genotype frequencies, as in the previous section.

This method of calculating genotype frequencies is used in a variety of different contexts. Box 10.3 describes a use in medical genetics.

BOX 10.3 DETERMINING CARRIER FREQUENCIES

Until recently, the Mediterranean island of Sardinia had one of the world's highest incidences of the severe recessive blood disorder, β-thalassemia, caused by mutation in the gene encoding β-globulin. In 1975, 1 in every 213 individuals was affected and in some isolated villages the incidence was as high as 1 in 100. As the first stage in a program aimed at reducing the incidence of the disease, the government wanted to know the level of carriers (or heterozygotes) in the population.

An incidence of 1 in 213 β-thalassemia sufferers represents a recessive phenotype frequency of 0.0047. Thus:

$q^2 = 0.0047$

$q = \sqrt{0.0047} = 0.0685$

$p = 1 - q = 1 - 0.0685 = 0.9315$

Carrier frequency = $2pq$ = 2 × 0.0685 × 0.9315 = 0.1276

This means that 12.76%, or 1 in 8, of the population were carrying the allele for β-thalassemia. In 1977, a public education program was initiated, using local meetings, posters, pamphlets, radio, television, and newspapers, and targeted at couples planning a pregnancy. The intention was to identify carriers and advise prospective parents of their chances of conceiving a child with β-thalassemia. One member of each couple was tested (a simple blood test could identify carriers). If this first result was positive, the second partner was tested. If both partners were found to be carriers, then they were offered prenatal diagnosis. In the first 3 years of the program, 694 couples were found to be at risk of conceiving an affected child. As the result of these couples' reproductive decisions, 42 fetuses were diagnosed as possessing the alleles for expression of β-thalassemia, of which 39 were electively aborted. By 1980, the incidence of β-thalassemia had dropped to 1 in 290. The screening program has continued and today the incidence stands at 1 in 4000.

Population screening for β-thalassemia carriers has become widespread in many countries where the disease is prevalent. Similar screening programs now occur in other populations at risk for other severe recessive disorders, such as the fatal Tay–Sachs disease among Ashkenazi Jews.

10.5   Testing for equilibrium

In describing methods for calculating allele and genotype frequencies it has been assumed that populations are in **genetic equilibrium** and that the frequencies calculated for one generation will be the same in succeeding generations. It is possible to test whether a population is in genetic equilibrium if heterozygotes can be identified as phenotypically distinct from the homozygous dominant individuals, as is, for example, possible with the MN blood groups. To test for equilibrium a comparison is made between the **observed** genotype frequencies and those **expected** if equilibrium exists. If the frequencies closely match, then equilibrium can be assumed to exist. If, however, there are large numerical discrepancies between observed and expected genotype frequencies, this suggests that a population may not be in equilibrium. A χ^2 statistical test (Section 4.6) should then be performed to determine whether the differences between observed and expected frequencies are significant. Box 10.4 shows the results of a χ^2 test on data obtained for the marsh frog, *Rana ridibunda*.

BOX 10.4 TESTING FOR EQUILIBRIUM

Genotypes formed from two alleles (**L** and **M**) of the gene for the enzyme lactate dehydrogenase were studied among 340 individuals of the marsh frog, *R. ridibunda*. The observed numbers of individuals were 133 **LL**, 135 **LM**, and 72 **MM**. A χ^2 test was used to assess whether this population was in equilibrium.

Calculation of observed allele frequencies:

Genotypes	Number of individuals	Number of L alleles	Number of M alleles
LL	133	266	
LM	135	135	135
MM	72		144
		401	279

Frequency of **L** = $\dfrac{401}{680}$ = 0.59

Frequency of **M** = $\dfrac{279}{680}$ = 0.41

From the observed allele frequencies, **expected genotype frequencies** are calculated using the Hardy–Weinberg frequencies of p^2, $2pq$, and q^2: **LL** = p^2, **LM** = $2pq$, and **MM** = q^2. Thus, **LL** = $(0.59)^2$ = 0.348, **LM** = 2 × 0.59 × 0.41 = 0.484, and **MM** = $(0.41)^2$ = 0.168.

A χ^2 test is used to assess whether there is any significant difference between the observed and expected frequencies:

Genotype	Observed numbers (O)	Expected numbers (E)	$O - E$	$(O - E)^2$	$(O - E)^2/E$
LL	133	118	15	225	1.91
LM	135	166	−31	961	5.79
MM	72	57	15	225	3.94
Total	340	340			11.67

- Null hypothesis: there is no difference between observed and expected genotype frequencies.

- Significance level = 0.05; degrees of freedom = 2 (number of possible genotypes − 1); calculated χ^2 value = 11.67; critical χ^2 value = 5.99

As the calculated χ^2 value is greater than the critical χ^2 value, the null hypothesis is rejected. There is, therefore, a significant difference between observed and expected genotype frequencies. The population is *not* in Hardy–Weinberg equilibrium.

10.6   Disturbances of the Hardy–Weinberg equilibrium

The Hardy–Weinberg equation enables us to predict allele and genotype frequencies in populations that are stable and unchanging. It is difficult, however, to find natural populations where all the assumptions required of the Hardy–Weinberg law rigorously exist. Rarely do we find large and randomly mating populations where no mutation, migration, or differential reproduction are occurring. Populations are dynamic – changes in size and structure are a natural part of their being. The final part of this chapter briefly considers the various factors that disturb equilibrium and how they might contribute to evolutionary change. We can think in terms of:

1. Factors that produce new variation (i.e. introduce new alleles into a population). **Mutation** is the main means of generating new alleles, although migration plays a subsidiary role.

2. Factors that change the profile of existing variation by altering allele frequencies within a population. **Migration** can have a major influence here along with genetic drift and differential reproduction.

10.7   Mutation

Within a diploid, sexually reproducing population the gene pool is reshuffled each generation to produce new combinations of genotypes among offspring. These new genotype combinations are the result of the processes of meiosis and the random fusion of gametes. They ensure variability among individuals.

These processes do not, however, produce any new variation. Only a mutation can create a new allele. Mutations are changes in genes and occur at random, without regard for any possible benefit or disadvantage to the organism. Thus, when considering the possible impact of mutations on a population it is important to know:

- How frequent a given mutation is
- The effect of the mutation – is it beneficial, harmful or neutral?
- How fast the new allele spreads among individuals of a population
- Whether the new allele is dominant or recessive in its effect on phenotypic expression

Mutation rates are generally expressed as the number of new alleles at a given locus per given number of gametes and are of the order of one new mutation per 100,000 gametes, or 1×10^{-5} gametes. Most new alleles are disadvantageous, reducing the fitness of an individual. It is believed, therefore, that over evolutionary time, mutation rates themselves have been subjected to selection pressures. Thus, the generally low mutation rate at most loci represents a balance between the potentially advantageous and disadvantageous effects of producing new alleles. Any one gamete is unlikely to have more than one new mutant allele.

With an average mutation rate of 1×10^{-5} per locus, even if the new allele conferred a selective advantage upon individuals, it would take many thousands of generations for the allele to reach significant levels within a population. Figure 10.1 shows, for example, that it would take of the order of 40,000 generations for a

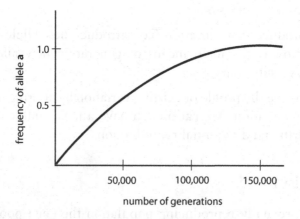

Figure 10.1 Changes in allele frequency by mutation alone. A population begins with one allele, **A**, at a locus. Allele **A** mutates to **a** at a frequency of 1×10^{-5}. The graph shows the change in frequency of allele **a** with time.

second favorable allele at a previously monomorphic locus to reach an allele frequency of 0.5! The process of mutation is an important evolutionary force within populations in that it is the only means of producing new alleles. It is not, however, a major force by itself in changing allele frequencies as mutation rates are too low. Other factors need to be operating within a population to bring about a significant increase in the frequency of a new allele. Migration and changing selective forces are examples of two such forces.

10.8 Migration

Plant and animal species frequently become divided into geographically separate populations. Migration occurs when individuals move between two such populations. If there is a large number of immigrants or a large difference in allele frequencies between the existing population and the immigrants, then a major change in allele frequency can occur in just one generation. Migration can be viewed as both a potentially positive and negative process. The positive aspect is its ability to introduce new alleles, and so novel variation, into a population. However, migration abolishes differences between populations. Thus, if there were major shifts in selective pressures this could mean no population had a suitable spectrum of variation to respond favorably.

Migration can be a one-off event or a continual process because, for example, two populations are no longer geographically isolated. In the latter case we consider migration more in terms of **gene flow**. Much gene flow has occurred, for example, over the last couple of centuries between black and white Americans. Most of the black population of the USA is descended from ancestors in West Africa where allele frequencies at many loci are different to those in the white Europeans who emigrated across the Atlantic. Gene flow, through intermarriage, is gradually reducing the differences at many loci. Annual gene mixing occurs between populations of the monarch butterfly, *Danaus plexippus*. The butterfly overwinters in central Mexico. In the spring, individuals begin to fly north towards Canada, mating and egg laying en route. Newly emerged adults join the northwards flight. As a result of this great communal migration there is little genetic differentiation between different populations.

Habitat destruction is a global conservation issue. It often fragments populations of plant or animal species already under threat of extinction. The result is small isolated populations that often show limited variability. In such cases gene flow is encouraged! Great efforts are directed towards maintaining habitat corridors between fragmented populations to encourage gene flow and, thus, maintenance of different alleles in the populations, so that, in effect, individuals are part of one large population and so avoid the problems of **genetic drift** and **inbreeding** associated with small populations.

10.9 Genetic drift: random changes in small populations

Allele frequencies in large populations may change from one generation to the next as the consequence of chance alone. However, any fluctuations in one generation are usually reversed in subsequent generations, so that no major changes occur in allele frequencies. The dynamics are very different in small populations. Consider Figure 10.2. Initially there was a pool of gametes in which 50% of alleles at a particular locus were **A** and 50% were **a**. After only two generations the proportion has shifted so that 80% of the alleles are **A** and only 20% are **a**! Such random fluctuations in allele frequencies are a characteristic of small populations. Their cumulative effect is known as **genetic drift**. No two populations will exhibit the same drift pattern. You can confirm this statement for yourself using colored beads to represent the gametes and repeating the events of Figure 10.2 five or 10 times, or by examining the hypothetical simulation represented in Figure 10.3.

Figure 10.3 illustrates another common feature of genetic drift – that it leads to the loss of genetic variation from a population. Alleles are eliminated or **fixed** (i.e. become the only allele present at a locus). As a result, individuals share fewer alleles and the population becomes **inbred**. Many loci are homozygous, with increased expression of deleterious recessive alleles. Survival rates are lowered and the population shrinks ever smaller, intensifying inbreeding effects. It can be clearly seen, therefore, that Hardy–Weinberg equilibrium does not exist in small populations.

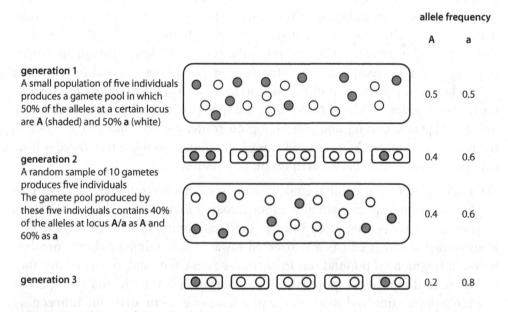

Figure 10.2 Genetic drift in small populations.

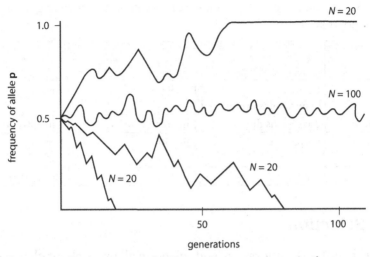

Figure 10.3 Simulating genetic drift. The graph represents the results of computer simulations of random changes in allele frequencies over 100 generations in three populations where $N = 20$ and in one population where $N = 100$. In all populations the initial allele frequencies of p and q were 0.5.

A population, though apparently stable and in genetic equilibrium, may show low levels of genetic variation because at some time in the past numbers were low. A **bottleneck** refers to a transitory but drastic shrinking of the population size, caused by a natural event, such as an earthquake, flood, drought, or disease epidemic. Population numbers generally recover after such circumstances; however, while they are low, drift can substantially alter allele frequencies so that subsequent populations show considerably less variation. A severe bottleneck is believed to have occurred about 10,000–12,000 years ago in the African cheetah, *Acinonyx jubatus*, population. Although numbers have grown again, it may be the reason for the extremely low level of genetic variation observed among cheetahs today. The low sperm counts and high infant mortality rates among captive populations are consequences of this genetic uniformity. Can the cheetah ultimately survive?

Related to a bottleneck is the **founder effect**, when a small group of individuals separate from the main large population and establish a colony in a new location. Again, numbers generally grow, but initially drift will be a major factor operating to change allele frequencies and reduce variation. Thus, in both bottleneck and founder situations, the group of surviving or founder individuals may possess different allele frequencies to the original population; for example, a higher incidence of a particular recessive allele which is then passed to many descendants. This situation can explain the above average frequencies of certain genetic diseases observed among isolated human populations or groups who have emigrated to new lands (Table 10.3).

TABLE 10.3 The founder effect: examples of elevated incidence of genetic diseases in particular populations

GROUP	DISORDER	FREQUENCY IN FOUNDER GROUP	MEAN WORLD FREQUENCY
Africaners	Familial hypercholesteremia	1 in 3000	1 in 1,000,000
Ashkenazi Jews	Tay–Sachs disease	1 in 3600	1 in 350,000
Bantu South Africans	Porphyria	1 in 250	1 in 25,000
Hopi Indians	Albinism	1 in 230	1 in 15,000

10.10 Selection

Genetic drift, mutation, migration, and selection all bring about changes in allele, and thus genotype, frequencies in populations. However, the changes caused by the first three factors are random. By contrast, the changes brought about under the influence of selection have a directional, or differential, aspect. As the result of selection, alleles favorable to the survival, and therefore reproduction, of an individual will increase in frequency in a population at the expense of alternatives.

In most populations, at any given moment in time, there are individuals with different alleles and therefore different genotypes. As a result of these inherent genetic differences, some of the individuals in a population will possess phenotypes that better adapt them to the prevailing environmental conditions. Such individuals have a **selective advantage**, and will survive and reproduce at the expense of others. Thus, certain genotypes, and therefore alleles, are being favored and, over time, allele frequencies will change in a population. If it is the conscious choice of humans that individuals survive, then we recognize the process as **artificial selection** (e.g. when agriculturists choose to breed from those individuals in a crop that show the most resistance to a certain pesticide or the selective breeding of domestic dogs over the past eight millennia). If selection occurs as a result of natural circumstances, then it is called **natural selection**.

Natural selection can be a major force in changing allele frequencies. In population genetics we attempt to quantify the strength of selection, and to calculate its effect on allele and genotype frequencies, and thus upon the variation shown by populations. Two key concepts used by population geneticists are **relative fitness** (w) and the **selection coefficient** (s). The first term is an indicator of the relative advantage of possessing certain genotypes, while the selection coefficient represents the downside of a given genotype! Thus, for a genotype to possess a fitness of 1 is good news, but if we are told $s = 1$ for a genotype, there is little hope of an individual possessing that genotype surviving long enough to reproduce.

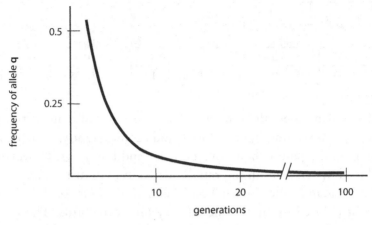

Figure 10.4 Changes in allele frequency under selection.

There are many questions that we can investigate if we know allele frequencies (i.e. p and q) and either s or w. At first sight the numerous mathematical models and formulas may seem daunting, but interesting insights into future population structure can be gained from their use. One issue is investigated in Figure 10.4. Suppose a recessive homozygote was lethal ($s = 1$). We might be interested to know how quickly the recessive allele would disappear from a population. The graph shows that initially there is a rapid drop in frequency of the recessive allele, as selection operates against the recessive homozygotes. Its frequency is halved after two generations and halved again after a further four. However, after that its rate of reduction declines. This is because many recessive alleles remain in the heterozygote, hidden to the force of selection that operates against phenotypes, and not alleles. A recessive allele only expresses itself in the phenotype of an individual as the recessive homozygote. As recessive alleles can be present in a population in the heterozygote members, it can be very difficult, indeed impossible, to completely eliminate alleles from a population – probably a good thing, in that it maintains variation in a population for "a rainy day." This notion can be illustrated by the following classic example of selection in a natural population.

For any starting frequency of the recessive allele the following formula enables us to calculate its frequency (q_n) after a predetermined number of generations and when $s = 1$ for the recessive homozygote:

$$q_n = \frac{q_o}{1 + nq_o}$$

where n is the number of generations since the original, q_o is the original frequency of q, and q_n is the frequency of q after n generations.

The following table shows the changes in recessive allele frequency over generations when the initial frequencies of both p and q are 0.5:

	Generation											
	0	1	2	3	4	5	6	10	20	40	70	100
p	0.50	0.67	0.75	0.80	0.83	0.86	0.88	0.91	0.95	0.98	0.99	0.99
q	0.50	0.33	0.25	0.20	0.17	0.14	0.12	0.09	0.05	0.02	0.01	0.01

Prior to the mid-nineteenth century, 99% of the population of the peppered moth, *Biston betularia*, was light colored and, consequently, well camouflaged when resting on the pale lichen-covered trees and buildings. However, as toxic gases produced by factories in the rapidly expanding industrial towns and cities of Victorian England killed the lichens and soot deposits darkened the buildings, the light-colored moth became an easy target for birds. The rare dark, or melanic, form of the moth suddenly gained a huge selective advantage as it was now the camouflaged variant. By 1900, it formed 95% of populations in industrial centers, although it remained rare in rural areas (Figure 10.5). Following the passing of laws in the 1950s and 1960s to restrict environmental pollution, lichens have recolonized formerly polluted areas and buildings are gradually being cleaned. The result is a steady decline in the melanic form, as the moth's habitat becomes pale colored once again. Similar examples of rapid changes in allele frequencies, and so resulting genotypes and phenotypes, include the rapid development of resistance to pesticides by insects and strains of the bacteria *Staphylococcus aureus* to the antibiotic methicillin – the MRSA strains.

Summary

- The goal of population genetics is to understand the genetic structure of a population, and the forces that determine and change its composition.

Figure 10.5 Light and dark forms of the peppered moth, *B. betularia*, at rest on a tree in an unpolluted area. Courtesy of Martinowks under CC BY-SA 3.0 license.

- The existence of various alleles at different loci produces genetic variation both between individuals within a population and between different populations.

- A fundamental measurement used in population genetics is the frequency with which different alleles occur at any given locus. For a gene with two alleles their frequency is represented by p and q, and $p + q = 1$.

- Knowing allele frequencies enables us to predict genotype frequencies in a population using the terms p^2, $2pq$, and q^2.

- The frequency of any allele in a population can be changed by recurrent mutation, migration, selection, or genetic drift.

- Natural selection is the most powerful force acting on a population's genetic composition, although in small populations genetic drift can also be an important factor.

- A randomly interbreeding population, in which no forces of change are acting, will show constant genotype frequencies for a given locus in successive generations. Such a population is said to be in genetic equilibrium and $p^2 + 2pq + q^2 = 1$ (the Hardy–Weinberg equation).

Problems

1. The frequency of a recessive allele in a large randomly mating population is 0.2. What is the frequency of:

   (a) The dominant allele?

   (b) The heterozygote?

2. In the Gaboon viper, *Bitis gabonica*, one main locus determines the lethality of its venom. Individuals with the genotype **W** are deadly poisonous; heterozygotes, those with **Vv**, are mildly poisonous, while any individuals homozygous recessive, **vv**, are non-poisonous. In a population of 3000 vipers the frequencies of the **V** and **v** alleles were found to be 0.75 and 0.25, respectively. How many of these vipers are non-poisonous?

3. In a large interbreeding population of Canadian hairless cats, 81% of individuals are homozygous for a recessive character. What percentage of individuals in the next generation would you expect to be:

   (a) Homozygous recessive?

   (b) Heterozygous?

   (c) homozygous dominant?

4. For each of the following, state whether it is an example of an allele, genotype, or phenotype frequency:

   (a) Approximately 1 in 2500 Caucasians are born with cystic fibrosis.

   (b) The percentage of carriers of the sickle cell allele in West Africa is approximately 13%.

   (c) The number of new mutations for achondroplasia is approximately 5 × 10^{-5}.

5. Hurler syndrome is an autosomal recessive disorder of mucopolysaccharide metabolism, resulting in short stature, mental retardation, and various bone malformations. The incidence of affected newborn infants is about 1 in 76,000. Assuming random mating, what is the frequency of heterozygotes?

6. A randomly mating population of sheep contains an autosomal recessive allele causing the coat to be curly.

   (a) If the incidence of curly-coated lambs is 10%, what is the frequency of the heterozygous carriers of the allele?

   (b) What is the frequency of heterozygotes among normal-fleeced sheep?

7. In certain grasses, the ability to grow in soils contaminated with the toxic metal nickel is determined by a dominant allele. If 78% of the seeds of the meadow grass, *Poa pratensis*, are able to germinate in contaminated soil, what is the frequency of the resistance allele?

8. (a) Name four evolutionary processes that can change allele frequencies in a population.

   (b) Recessive alleles are often lethal when homozygous and so there is selection in each generation against such alleles. Yet such alleles are rarely completely eliminated from a population. Why?

   (c) How does the frequency of heterozygotes in an inbred population compare with that in a randomly mating population with the same allele frequency?

   (d) What does it mean when we say an allele is fixed?

   (e) Why is mutation a weak force for changing allele frequencies?

   (f) What factors cause genetic drift?

9. In a certain population of Australian banjo frogs, 120 are green, 60 are brownish-green, and 20 are brown. The allele for brown is denoted **B** and that for green **G**. The two alleles show incomplete dominance.

   (a) What are the genotype frequencies in the population?

   (b) What are the allele frequencies?

(c) What are the expected frequencies of the genotypes if the population is at Hardy–Weinberg equilibrium?

(d) Is the population in Hardy–Weinberg equilibrium?

10. Seventy tiger salamanders from one pond in west Texas were examined for genetic variation at the enzyme locus malate dehydrogenase. Two alleles, **A** and **B**, were identified using gel electrophoresis. Among the salamanders the distribution of genotypes was:

Enzyme genotype	Number of individuals
AA	15
AB	49
BB	6

Is this population in Hardy–Weinberg equilibrium?

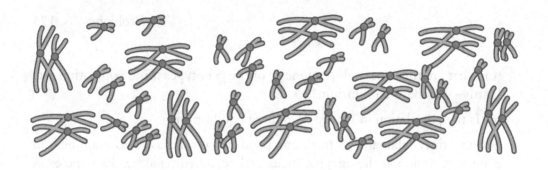

Heredity at the Molecular Level

In August 1953, the scientific community was thrown into a fever of excitement by a short letter that appeared in the leading scientific journal *Nature*. The authors of the letter, Francis Crick (1916–2004) and James Watson (*b.* 1928), proudly announced the discovery of the structure of the molecule of inheritance – DNA. Ten years later Crick and Watson, along with Maurice Wilkins, would receive the Nobel Prize in Physiology or Medicine for their insights. To understand the structure of DNA had been the goal of many biologists for much of the previous decade; certainly since 1944, when Oswald Avery (1877–1955) and his coworkers had convincingly shown that DNA was the critical component of the bacterium *Streptococcus pneumoniae*, capable of transforming harmless avirulent varieties into pneumonia-provoking versions. From that moment the race was on to elucidate DNA's molecular structure. Crick and Watson reached the winning line first. The model they proposed was supported by all the available evidence and, as they were quick to appreciate, suggested a mechanism whereby the DNA molecule could be perfectly copied – a crucial requirement of hereditary biological material.

This chapter describes:

- The basic structure of DNA
- The packaging of DNA to form chromatin
- The mechanism of DNA replication

11.1   The structure of DNA – a summary

DNA is a **polymer**, which means, biochemically speaking, it is a huge chain-like molecule whose great size is a result of endless repetition of a much smaller molecule or **monomer**. The repeating monomer of DNA is the **nucleotide**, millions of which join end-to-end to produce long **polynucleotide** chains. Two of these chains spiral around each other to produce the famous "double-helical" form of DNA.

11.2   The biochemical nature of nucleotides

Each nucleotide in DNA possesses **three** components:

- A **phosphate group**
- A sugar (**deoxyribose**)
- A **nitrogenous base**

Deoxyribose belongs to a group of sugars known as **pentoses** because each sugar contains five carbon atoms. Pentoses exist either as straight chains or pentagon rings. The ring form is found in a nucleotide (Figure 11.1). Only one carbon atom is shown in the diagram; the other four are at the angles of the pentagon. Furthermore, starting with the right-hand carbon, each is numbered from $1'$ to $5'$ (the numbers and the dash, or "prime," aid in communicating biochemical structure and function). The nucleotides of DNA are called **deoxyribonucleotides**, because of the sugar they contain. However, this cumbersome term is generally simplified to **nucleotide**; although, because of the presence of deoxyribose, the polymer of nucleotides we are describing is formally known as **deoxyribonucleic acid**, abbreviated to **DNA**.

The **nitrogenous bases** (so called because they contain nitrogen and combine with hydrogen ions in acidic solution) vary within a nucleotide. There are four main variants to be found in DNA: **adenine**, **guanine**, **cytosine**, and **thymine** (commonly represented by their first letters: **A**, **G**, **C**, and **T**). As Figure 11.1

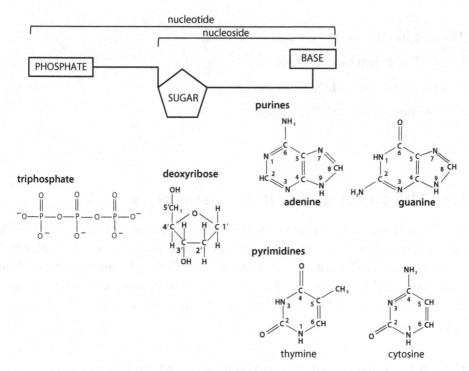

Figure 11.1 The components of a deoxyribonucleotide.

shows, the four bases differ in their complexity. Cytosine and thymine each possess a single carbon–nitrogen ring and are called **pyrimidines**, while adenine and guanine possess a double ring and are referred to as **purines**.

Within each nucleotide, the base and phosphoric acid are bonded to the sugar (Figure 11.1). The base links to the 1′ carbon. The term **nucleoside** refers to a base bonded with a pentose sugar. The phosphoric acid links to the 5′ carbon of the sugar (Figure 11.1).

11.3 The double helix

Crick and Watson's key contribution towards understanding the nature of DNA was to show how the nucleotides are organized within the molecule. Analysis of all currently available evidence led them to propose the **double-helix** model, in which DNA is envisaged as a twisted ladder with bonded bases as its rungs (Figure 11.2). Alternating sugars and phosphate groups constitute the two sides of the ladder, referred to as the **sugar-phosphate backbone**. The phosphate of one nucleotide links to the 3′ carbon of the sugar of the adjacent nucleotide by a **phosphodiester** bond. The bases project from this sugar-phosphate backbone at regular intervals at an angle of approximately 90° (Figure 11.2).

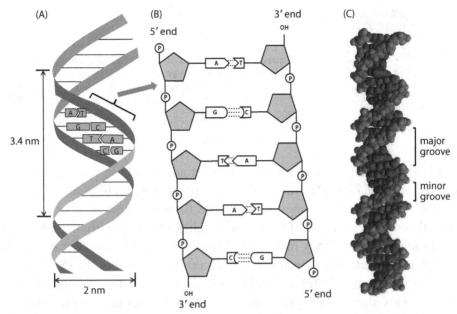

Figure 11.2 The Watson–Crick model of DNA. (A) A simplified representation of DNA as a double helix. (B) A short length of DNA untwisted to show the relative positions of the deoxyribose (the pentagons), phosphate (circles), and bases. Note also the anti-parallel nature of the two strands. (C) A space-filling model of the B form of DNA showing major and minor grooves.

Cross-chain binding occurs between these projecting bases and is highly specific. A purine can only bind with a pyrimidine. In fact, adenine must link with thymine and guanine must link with cytosine – the famous **A–T** and **C–G** pairs of DNA. This specific, complementary pairing is the key to understanding much about the structure of DNA, its replication, and accessing the stored hereditary information for use by the cell. Regarding the structure of the DNA molecule, it is only the pairing of A with T and C with G that orientates the bases in such a way as to produce:

- Weak attractive forces between opposing bases that can hold the two polynucleotide chains together. These forces are known as **hydrogen bonds**. Two hydrogen bonds link A and T, while G and C are linked by three hydrogen bonds.

- A regularly structured molecule – another characteristic feature of the DNA molecule, which was very clearly indicated in the X-ray diffraction pictures of DNA (Figure 11.3) that Rosalind Franklin (1920–1958) produced in the early 1950s.

The correct orientation for hydrogen bonding only occurs when the two polynucleotide chains are orientated in opposite directions (i.e. are **anti-parallel**) (Figure 11.2).

Figure 11.3 X-ray diffraction image of a DNA molecule. The technique involves firing a beam of X-rays at a crystal. The atoms in the DNA deflect the X-rays. The scattered X-rays are caught on a photographic plate, producing a pattern that can be used to deduce the arrangement of atoms in the molecule.

Each base pair is slightly offset relative to its adjacent base pairs. This explains the twist or spiral of a DNA molecule (i.e. why the polynucleotide chains assume a double helix). In its commonest form, the **B form**, one complete twist of the molecule occurs every 10–10.5 nucleotide pairs, producing predictable and regular dimensions to a DNA molecule. There are just 0.34 nm between adjacent bases; one complete spiral occurs every 3.4 nm and the width of a DNA molecule is a uniform 2 nm (Figure 11.2). The spiraling of the two nucleotide strands produces two alternating spiral grooves referred to as the **major groove** and the **minor groove** (Figure 11.2). These grooves are important binding sites for proteins regulating gene expression. In areas of active transcription, and where there are stretches of CG sequences, DNA is a more open, less regular helix, the **Z form**, while in dehydrated samples the DNA compacts to the **A form**.

11.4   DNA packaging

When describing the structure of DNA we tend to depict the double helix continuing in a long straight line. If, in reality, the approximately 3 billion (3×10^9) nucleotide pairs of the DNA in each human cell were stretched out in this way, it would extend for 2 m! Yet, this 2 m of DNA is packaged into a tiny cell nucleus of approximately 10 μm diameter. This packaging is possible because an organized system of coiling eukaryotic DNA occurs, which reduces its linear length 10,000-fold.

To achieve this condensation the DNA first wraps around groups of **histone** proteins forming **nucleosomes** (Figure 11.4). Between 145 and 147 nucleotides are wound around each histone core, while 20–60 nucleotides form a spacer region before the next nucleosome. The result is a beads-on-a-string effect (Figure 11.4). The nucleosome string, often called a 10-nm fiber, is further folded and compacted into a thicker fiber, called a **solenoid** or 30-nm fiber (Figure 11.5). This and the previous, more extended nucleosome arrangement represent the level of organization of the DNA during normal cell functioning, referred to as **chromatin**. To achieve the additional size reduction needed to produce the chromosomes visible during mitosis and meiosis, the chromatin fibers fold again, forming supercoiled loops radiating from a protein core (Figure 11.5).

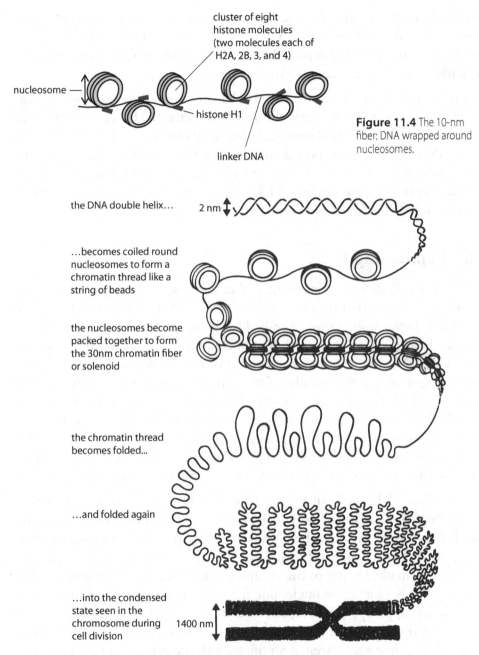

cluster of eight
histone molecules
(two molecules each of
H2A, 2B, 3, and 4)

nucleosome —

histone H1

linker DNA

Figure 11.4 The 10-nm fiber: DNA wrapped around nucleosomes.

the DNA double helix…   2 nm

…becomes coiled round nucleosomes to form a chromatin thread like a string of beads

the nucleosomes become packed together to form the 30nm chromatin fiber or solenoid

the chromatin thread becomes folded…

…and folded again

…into the condensed state seen in the chromosome during cell division   1400 nm

Figure 11.5 The different levels of organization of DNA in a chromosome.

There are eight histone molecules in a nucleosome – two each of histones 2A, 2B, 3, and 4. All histone proteins possess many basic amino acids (i.e. many lysine and arginine molecules) so that the histones have a positive charge to attract the negatively charged DNA. The extended 10-nm chromatin fiber organization is also called **euchromatin** and is the form of DNA when it is being actively transcribed. A fifth histone, H1, helps to compact the 10-nm chromatin fiber into the 30-nm fiber. It binds to the nucleosome at the point of exit of the DNA molecule (Figure 11.4). A compacted 30-nm fiber is also referred to as **hetero-chromatin** in which genes are inactive.

The single small circular DNA molecule of bacterial cells does not need packaging to the same extent as eukaryotic chromosomes. Bacterial DNA is **supercoiled** to compact it into the nucleoid region of a bacterial cell. Supercoils are produced when a DNA helix is over-rotated (imagine twisting a rubber band until the original coils fold over one another to form a condensed mass). **Topoisomerases** create the supercoils by transiently nicking DNA, rotating the cut ends around each other and then resealing the cut ends.

11.5 Replicating DNA

Every time a cell divides it is essential that each new daughter cell receives two complete copies (i.e. two sets of chromosomes) of the organism's biological information. As this information is embedded within the molecular structure of the double helix (i.e. it is the sequence of nucleotides), this means that extensive and accurate DNA replication must occur prior to each cell division. Crick and Watson recognized that their model of the structure of DNA also indicated how the molecule could be accurately replicated; as they tentatively suggested in their landmark 1953 *Nature* paper: "It has not escaped our notice that the specific pairing we have postulated immediately suggests a possible copying mechanism for the genetic material." The key to the replication of DNA resides in the complementary pairing of the bases – adenine with thymine and guanine with cytosine.

Crick and Watson's ideas about the mechanism of DNA replication went something like this. Suppose the two polynucleotide strands separate (this is chemically plausible because the hydrogen bonds holding the bases, and so strands, together are very weak). Each separated strand could then act as a template for a mirror copy to be made. Owing to the complementary nature of the base pairing, wherever there is an adenine nucleotide in the template strand only a thymine nucleotide can align opposite and, likewise, only guanine and cytosine nucleotides could pair with each other. In this way a new complementary strand could be assembled against each parental template, with the result that two molecules identical to the original molecule are produced (Figure 11.6). It took scientists a dozen or so years after the publication of Crick and Watson's model to confirm their mechanism of DNA replication and to work out the molecular details. Matthew Meselson (*b.* 1930) and Frank Stahl's (*b.* 1929) elegant experiments were crucial in confirming this overall **semi-conservative** pattern of replication (Box 11.1). Arthur Kornberg (1918–2007) and his co-workers worked out the key molecular details (Section 11.6).

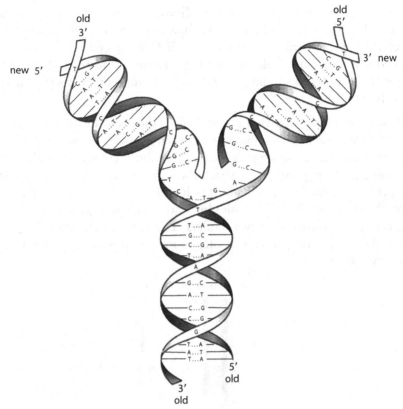

Figure 11.6 DNA replication. A section of the double helix (termed "old") unwinds to produce two separate template strands. By complementary base pairing, two new strands are produced.

BOX 11.1 SEMI-CONSERVATIVE DNA REPLICATION

The term "semi-conservative" refers to the fact that any newly produced DNA molecule contains one original parental strand and one newly synthesized daughter strand. Although Watson and Crick's semi-conservative scheme seemed the most likely, other patterns were suggested in the 1950s, as illustrated below, such as "conservative" replication, during which one daughter molecule retained both parental strands and the other had two newly synthesized strands, or "dispersed," in which parental and newly synthesized strands were randomly found in daughter molecules.

In one of the classic experiments of molecular biology, Meselson and Stahl devised a strategy that could distinguish between the three proposed mechanisms. They differentially labeled parental and daughter chains. To do this they used different atomic forms or **isotopes** of nitrogen – the normal ^{14}N type and the heavier ^{15}N form. It is possible to control which isotope the nitrogen-containing bases of DNA possess by limiting which form is available in an experimental organism's culture medium. DNA molecules containing one or other or a mixture of the two isotopes will be of slightly different mass and can be separated and identified by **density gradient centrifugation**. Centrifugation separates cellular constituents according to their mass. If DNA is centrifuged through a test tube containing cesium chloride, the latter forms a density gradient and the different DNA molecules sediment or "band" at different, characteristic positions, according to their mass.

In their experiment, Meselson and Stahl used the bacterium *Escherichia coli*. Their results were entirely in agreement with a semi-conservative mode of DNA replication. Initially, *E. coli* were grown in a medium containing only ^{15}N. At the beginning of the experiment the bacteria were transferred to ^{14}N medium. After one generation the DNA molecules sedimented at a site intermediate between pure ^{15}N and pure ^{14}N. This result eliminated the conservative model, but could support either of the other two models. The result, an intermediate band and one at the ^{14}N position, after a further round could only support the semi-conservative model.

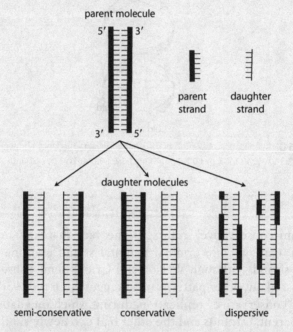

Box 11.1 Figure 1 Three suggested patterns of DNA replication.

11.6   Details of DNA replication

A large number of enzymes and other proteins are employed to ensure accurate DNA replication. For example, before replication can commence the double helix must first be unwound, followed by separation or "unzipping," of the two strands. **Helicases** break the hydrogen bonds between the bases on opposite strands of the parental DNA molecule and thus unwind the double helix. **Topoisomerases** remove the supercoils that then build up ahead of the unwinding DNA. They achieve this by causing transient breakages in the polynucleotide chains; either in one (type I enzymes) or both strands (type II enzymes). **Binding proteins** coat the separated single strands to prevent them re-annealing.

Once single template strands are exposed, **DNA polymerases** align the correct nucleotides opposite their complementary bases and catalyze the formation of new phosphodiester bonds between adjacent nucleotides. This complementary copying occurs in a different way on the two template strands because DNA polymerases can only synthesize DNA in a $5' \rightarrow 3'$ direction. This means the template must be read in a $3' \rightarrow 5'$ direction. However, the two strands of DNA are orientated in an anti-parallel fashion. Figure 11.7 illustrates the solution to this problem. Only one strand can be synthesized continuously – that copied from the **leading strand**, which is read in a $3' \rightarrow 5'$ direction. In order for the DNA polymerase to be able to synthesize in a $5' \rightarrow 3'$ direction against the other template strand, replication proceeds in the apparently wrong direction in small sections. The resulting **Okazaki** fragments (named after the Japanese biologist Reiji Okazaki who first demonstrated their existence) are eventually joined together by another enzyme – **DNA ligase** (Figure 11.7). As the new DNA molecules grow they wind around each other to form double helices. Note also in Figure 11.7 that each Okazaki fragment starts with a short sequence of RNA! This is because DNA polymerases cannot start DNA synthesis *de novo*. Thus, during DNA replication a **primase** produces a short "starter" sequence of RNA to which a DNA polymerase adds deoxyribonucleotides. Later in DNA replication the RNA is removed and replaced by DNA.

The process of continuous replication of one parental strand and discontinuous replication of the other continues until the entire DNA molecule has been

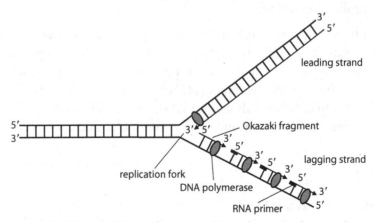

Figure 11.7 The role of Okazaki fragments in DNA replication. Leading-strand replication is continuous, while lagging-strand replication is discontinuous via Okazaki fragments.

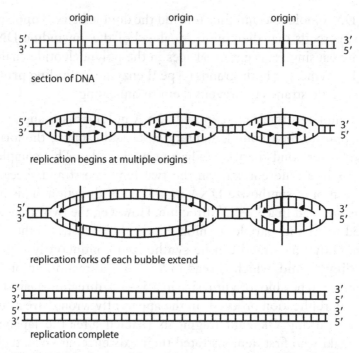

Figure 11.8 Replication origins and their movement along a DNA molecule.

duplicated (or at least a section). In most organisms, DNA replication would be an extraordinarily time-consuming process if it proceeded linearly from one end of a chromosome to the other. For example, it would take 2 months for an average-sized human chromosome to be replicated! Instead, replication begins simultaneously at many different sites along a eukaryotic chromosome, called **replication origins** (Figure 11.8). Eventually the many separately replicated sections join together. In the single small circular chromosome of bacteria, there is a single replication origin.

In addition to adding new nucleotides, most DNA polymerases also **proofread** the newly synthesized DNA. If an incorrect nucleotide has been inserted, these enzymes can recognize the mismatched pair, excise the wrong nucleotide, and replace it with the correct one. Hence, DNA replication is a virtually error-free process; for example, DNA polymerase III of the bacterium *E. coli* is believed to leave an uncorrected error just once in 5×10^9 nucleotides.

Summary

- The genetic material is DNA.

- DNA consists of two anti-parallel polynucleotide chains held together in a double helix by hydrogen bonds between the chains.

- Each nucleotide consists of a sugar (deoxyribose), a phosphate, and a nitrogenous base.

- Purines and pyrimidines are the two types of bases. Adenine and guanine are purines, while cytosine and thymine are pyrimidines.

- Complementary pairing of bases occurs across chains: adenine with thymine and guanine with cytosine.

- Exact replication of a DNA molecule occurs in a semi-conservative fashion. Each separated parental strand acts as a template for a complementary copy to be synthesized.

- Owing to the anti-parallel nature of the double helix, DNA is synthesized continuously only on one strand. On the opposite strand, synthesis produces short fragments that are later joined.

- DNA polymerase is the main DNA synthesis enzyme. It can also detect and correct misincorporated errors.

Problems

1. In a fragment of double-stranded DNA there are a total of 200 base pairs, of which 45 are thymine. How many of the following are there?

   (a) Nucleotides.

   (b) Complementary base pairs.

   (c) Adenine molecules.

   (d) Deoxyribose molecules.

   (e) Cytosine molecules.

2. A single strand of DNA contains the base sequence 5′-ACCGGTAGAATCG-3′. A complementary strand is synthesized from this template strand.

   (a) What is the sequence of the new strand?

   (b) In which direction will the DNA polymerase move along the template strand?

3. If 27% of the bases of an organism's DNA are guanine, what percentage are adenine?

4. In DNA, does the proper alignment of bases needed for hydrogen bonding of complementary bases occur when the two polynucleotide chains are orientated in the same direction or opposite directions?

5. The double-stranded DNA molecules extracted from a newly discovered virus were found to be 102 μm in length.

   (a) How many complete turns of the two chains are present?

   (b) How many nucleotide pairs are present in one such molecule?

6. For double-stranded DNA which of the following base ratios are always equal to 1?

   (a) A + T/G + C.

   (b) C/G.

   (c) T + G/C + A.

   (d) A/G.

7. Indicate whether each of the following statements are true or false.

   (a) A + T = G + C.

   (b) The two strands of a DNA double helix are identical.

   (c) If the base sequence on one DNA strand is known, the sequence of the second can be deduced.

   (d) The structure of DNA is invariant.

   (e) lf there are 34% adenine bases in a DNA molecule, then there will also be 34% cytosine bases.

8. What is meant by the terms semi-conservative and conservative replication?

9. Why is DNA replication continuous alongside one template strand and discontinuous alongside the other?

10. If 39% of the DNA for the slime mold, *Dictyostelium discoidium*, are adenine, what can you conclude about the base composition of the DNA of this species?

11. The double-stranded DNA molecule of the Epstein–Barr virus contains 172,280 base pairs.

   (a) How many nucleotides are present?

   (b) What would be the length of this molecule?

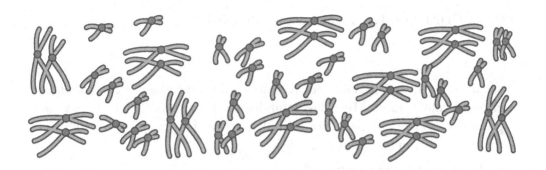

From Genes to Proteins

Why should some tomato plants produce red fruit and other plants produce yellow fruit? The explanation offered in Chapter 2 for these different phenotypes referred to the different fruit color alleles the two types of plants possessed, but was not particularly concerned with considering the nature of the different information the two types of alleles contained and how that influenced phenotype. The focus of Chapter 2 was, after all, a discussion of inheritance patterns and ways of predicting what color fruit future generations of plants might produce. This and the previous chapter are, however, concerned with different issues. They address the nature of the information contained within individual genes, and how it is accessed and used by a cell. Chapter 11 described the molecular structure of a gene's key constituent (i.e. DNA) and how it can be accurately copied each time a new cell is produced. This chapter considers the expression of the information contained within a gene.

The key molecule in this chapter's story is protein (Box 12.1). Most genes contain the information for the production of a specific protein, which leads directly or indirectly to expression of a particular phenotype. Molecular biologists have worked out how the DNA of genes codes for proteins and can describe, in considerable detail, how a cell produces a protein from the encoded information. Much less is understood, however, about the way a given gene's protein results in a final phenotype. This chapter will address these three key areas of gene expression:

- The nature of the genetic code

- The flow of information from DNA to protein

- The expression of different phenotypes

BOX 12.1 THE NATURE OF A PROTEIN MOLECULE

Proteins are polymers consisting of chains of amino acids linked by peptide bonds. Twenty different amino acids are found within proteins. Each different protein has a different specific amino acid sequence. As a single protein commonly consists of several hundred, often thousands, of amino acids, and at each site in the chain there are 20 possibilities, the variety of proteins is virtually limitless!

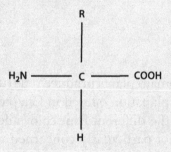

H = hydrogen atom
COOH = carboxyl group
NH_2 = amine group
R is different for each amino acid

Box 12.1 Figure 1 Generalized structure of an amino acid.

The terms **protein** and **polypeptide** are often used interchangeably, but they in fact refer to different aspects of protein structure. A polypeptide is a single chain of amino acids. The term protein should be reserved for the functional molecule. Often a single polypeptide chain is the functional protein. In such cases protein and polypeptide can be used synonymously. Other proteins, however, consist of two or more polypeptide chains. The two terms, then, refer to different states.

The chain of peptide-bonded amino acids represents the first level, or **primary structure**, of a protein. The chain assumes other specific forms before the polypeptide is functional. The **secondary structure** results when chains spiral to produce an α-**helix** or different regions associate forming a β-**sheet**. Some proteins, often those with a structural role, are functional at this secondary level. The functions of many other proteins (for example, enzymes) are dependent upon a specific three-dimensional shape that results when chains in their secondary state fold, producing the **tertiary structure**.

Secondary and tertiary structures are stabilized by a variety of interactions between the amino acids, including disulfide, hydrogen, and ionic bonding. As stated earlier, two or more polypeptide chains sometimes associate to produce a functional protein. The multisubunit form of a protein represents its **quaternary structure**.

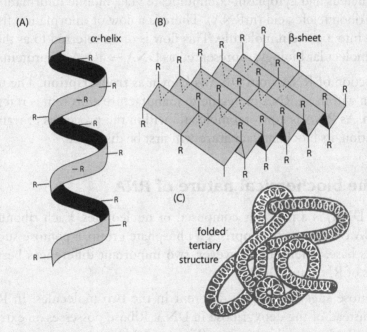

Box 12.1 Figure 2 Secondary (A and B) and tertiary (C) protein structures.

12.1   Information flow from gene to protein – a summary

A gene comprises a linear sequence of nucleotide pairs, while a protein is a linear sequence of amino acids. There is a direct and specific relationship between the two sequences: the nucleotide sequence determines the amino acid sequence. Each consecutive three nucleotides specifies a single amino acid of the encoded protein. This **collinearity** of sequences between a gene and a protein is illustrated in Figure 12.1, and is discussed in more detail in Section 12.6.

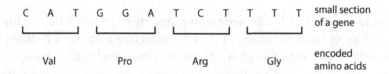

Figure 12.1 Collinearity between gene and encoded protein. Each consecutive three nucleotides encodes an amino acid. Val, valine; Pro, proline; Arg, arginine; Gly, glycine.

The encoded information and the site of protein synthesis are, however, physically separated. Synthesis of proteins occurs at ribosomes within the cell cytoplasm, yet the instructions for their synthesis reside within DNA in the cell nucleus. Nucleus and cytoplasm communicate via a mobile information carrier, messenger ribonucleic acid (mRNA). There is a flow of information from DNA via mRNA into a protein molecule. This flow is often referred to as the **central dogma** of molecular biology, represented as DNA → RNA → protein.

The production of RNA from DNA is known as **transcription**. The use of the information within an RNA molecule to manufacture a protein is referred to as **translation**. As RNA is a pivotal molecule within the processes of transcription and translation, its biochemical nature will first be discussed.

12.2   The biochemical nature of RNA

RNA, like DNA, is a polymer composed of nucleotides. Each ribonucleotide, like its DNA counterpart, comprises a phosphate group, a pentose sugar, and a nitrogenous base. There are, however, two important differences between the nucleotides of RNA and DNA:

1. The pentose sugar is slightly different in the two molecules. In RNA it is ribose, instead of the deoxyribose in DNA. Ribose possesses an extra oxygen atom attached to the third carbon (Figure 12.2A).

2. As in DNA, there are four different nitrogenous bases found within RNA. Three are the same in the two molecules (i.e. adenine, guanine, and cytosine). However, instead of thymine, RNA contains **uracil**. Uracil is structurally similar to thymine and can pair with adenine. It possesses a single hydrogen atom, instead of a methyl group (CH_3) on carbon 5 (Figure 12.2B).

Phosphodiester bonds link adjacent nucleotides in RNA, as in DNA. However, while DNA is double stranded, RNA usually only occurs as a single-stranded chain. Furthermore, because each RNA molecule contains the information relevant to just one gene, it is much shorter than a DNA molecule – thousands, rather than millions, of nucleotides in length.

OH

5′CH₂   OH
                O
4′C H   H C 1′
   3′   C—C   2′
   H   H
      OH   OH

(A) ribose

O
‖
C
HN 4   CH
  3   5
   C 2 1 6 CH
O ‖   N
        H

(B) uracil

Figure 12.2 Biochemical structures of (A) ribose and (B) uracil.

12.3 Producing RNA: transcription

Transcription is the process by which an RNA molecule is formed in a similar manner to DNA replication (i.e. a DNA strand is used as a template for synthesis of a complementary molecule). The process is illustrated diagrammatically in Figure 12.3. As in DNA replication, the DNA unwinds, hydrogen bonds between complementary base pairs of a gene are broken and the two polynucleotide strands separate. In contrast to DNA replication, only one of the two DNA strands is a template for RNA synthesis. Free ribonucleotides pair with the exposed bases of the template strand and the sugar and phosphate groups bond, producing a new single-stranded RNA molecule, which leaves the nucleus via a pore in the envelope. The transcribed RNA strand will, therefore, possess an identical sequence (except for the substitution of uracil for thymine nucleotide) to the DNA strand that did *not* serve as the template, which is termed the **sense strand** (Figure 12.3). The template strand is the **antisense strand**.

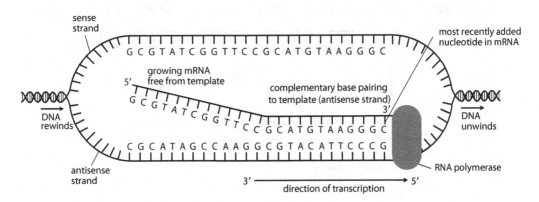

Figure 12.3 A summary of transcription. As DNA unwinds, transcription occurs from 3′ to 5′ using the antisense strand as a template.

As only a short region of the DNA molecule, a gene, is transcribed into an RNA molecule, this raises the question as to how the enzymes know where to begin and end RNA synthesis, or indeed which genes should be transcribed and when. As during DNA replication, a polymerase, this time **RNA polymerase**, is the principal transcription enzyme. RNA polymerase recognizes and binds to a special start or **promoter** sequence – a short distance in front of, or **upstream** from the gene to be transcribed. Once bound, the polymerase moves along the template strand in a $5' \rightarrow 3'$ direction, catalyzing polymerization of the ribonucleotides, until a termination sequence is reached which signals the end of transcription (and also the addition of a polyA tail, i.e. a stretch of up to 250 adenine nucleotides). This, and a special guanine cap at the beginning of the RNA transcript, are believed to protect the new transcript from degradation by cellular nucleases.

12.4   The different forms of RNA

All RNA molecules are produced in the same way, by transcription, from a DNA template and they all possess the same basic structure, as described in Section 12.2. They do not, however, all function in the same way. There are three main types of RNA, each with a different role in the cell. These three types are **messenger RNA (mRNA)**, **transfer RNA (tRNA)**, and **ribosomal RNA (rRNA)**. **mRNA** is the information carrier. Its nucleotide sequence represents the instructions for assembly of a precise amino acid sequence. It is the intermediary in the flow of information from DNA to protein. With regard to the other two types, the RNA molecule is the final product of gene expression (Figure 12.4). **tRNA** brings amino acids to the mRNA so they can be incorporated into a growing polypeptide chain. **rRNA** is a key component of ribosomes – the protein synthesis factory. Other types of RNA have recently been discovered with a variety of structural, regulatory, and enzymatic functions, such as small nuclear RNA that removes introns from pre-mRNA (see Section 12.5). In prokaryotes, all types of RNA are produced by the same RNA polymerase, while eukaryotes possess

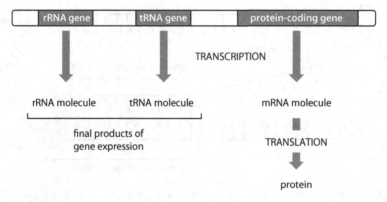

Figure 12.4 The three major types of RNA produced by transcription.

three different RNA polymerases. RNA polymerase I transcribes rRNA, mRNA is transcribed by RNA polymerase II, and RNA polymerase III produces tRNA.

Transcription produces an extended single-stranded molecule. Only mRNA is functional in this state. The transcribed single strands of tRNA and rRNA fold into characteristic three-dimensional forms of which different regions mediate different aspects of translation, such as the anticodon region of tRNA which recognizes and binds to mRNA codons (Section 12.7). Although mRNA is functional in an unfolded single-stranded state, in eukaryotes this is only after sections of the original transcript have been removed (Section 12.5), and the mRNA molecule has been **capped** at both the 5′ and 3′ ends:

An extra methylated guanine ribonucleotide is added at the 5′ end of the newly synthesized mRNA, which signals the beginning of the mRNA to ribosomes during translation.

Approximately 250 adenine ribonucleotides are added to the 3′ end of the newly synthesized mRNA. This **polyA tail** also functions in ribosome recognition of the mRNA molecule and helps to protect mRNA from exonuclease degradation.

12.5 Exons and introns

In the mid-1970s, molecular biologists were surprised to discover that the information within most eukaryotic genes is not continuous, but broken into coding and non-coding regions. The sections containing the biological information became known as **exons**, and the intervening segments as **introns**. The RNA molecule produced by transcription, referred to as **pre-mRNA**, includes both introns and exons. Thus, before mRNA can be used to direct the synthesis of proteins, the introns have to be removed and the remaining coding sections **spliced** together. Specific signal sequences at the boundaries of exons and introns signal splice sites to the **spliceosome** – a large complex of almost 300 proteins and different types of **small nuclear RNA** (**snRNA**). Splicing is a potentially risky process. If an intron is not precisely removed then, quite obviously, the mRNA sequence is changed, with the consequence that the encoded protein may not function as well or at all (Figure 12.5).

It is unclear why the important hereditary information should be broken into exons, sometimes a great many; for example, the human dystrophin gene (which causes the degenerative muscular disorder, muscular dystrophy, when faulty) has 79 exons. Some interesting examples of ways in which the presence of introns appear to maximize the use of a gene's base sequence have emerged. For example, some genes contain alternative splice sites. These allow the same primary transcript to be spliced in different ways, ultimately producing different protein products from the same transcript: the human thyroid hormone calcitonin and

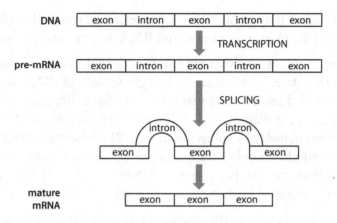

Figure 12.5 Stages in production of mature mRNA.

the brain calcitonin gene-related peptide are products of the same gene that has been differently spliced in the thyroid gland and the brain. There are several cases of separate genes encoded within the introns of large genes. For example, both the human neurofibromatosis and clotting factor genes contain other, unrelated, genes within their introns!

Other, evolutionary based, explanations have been presented for the existence of split genes. Attention has, for example, been drawn to the fact that, in at least some proteins, each exon codes for a different functional subcomponent. This observation has prompted the suggestion that during evolution exons from different genes can recombine to produce new, advantageous, combinations of biological information. Such a process is potentially more likely to produce new functional proteins than random arrangements of a gene. However, whatever the reason for the evolution of split genes, the advantages must outweigh the potential harm that can result from inaccurate splicing.

12.6   The genetic code

In the next section we consider details of the mechanism by which the information encoded within the nucleotide sequence of mRNA is translated into a protein of a specific amino acid sequence. First, however, we should consider the nature of the code. This section examines the main properties of the **genetic code**.

The code is a **triplet** (three-letter) code. Each consecutive three bases, known as a **codon**, represents an amino acid (Figure 12.1). There are 64 different ways the four DNA bases can be organized into triplets. Thus, there are 64 codons. Yet there are only 20 different amino acids found in proteins. This means that most amino acids are represented by more than one codon (i.e. the code is **degenerate**). Molecular biologists realized in the early 1960s, when much effort was being devoted to cracking the genetic code, that the code had to be degenerate because a triplet code represented the minimum number of bases that can encode all 20 amino acids. A code based on one base per amino acid gives only four variants, while couplets, e.g. G followed by A, or C and G, can only produce 16 variants. Table 12.1 shows how organisms deal with the excess codons. The different codons for a given amino acid are generally closely related. It is only the last base of the triplet that varies between codons; see, for example, arginine or lysine. Three codons, the **stop** codons, do not represent amino acids. Instead, they signal the end of a protein during translation. There is a universal **start** codon, AUG, which signals the start of translation. The amino acid, methionine, encoded by AUG is often removed from the newly synthesized polypeptide.

The code, as outlined in Table 12.1, is universal – it applies to all living organisms. One exception is mitochondria, which have their own extranuclear DNA (Box 12.2). In the human mitochondrial genome there are only two stop codons, UAA and UAG. The codon UGA, normally a stop codon, instead encodes tryptophan. The general universality of the genetic code represents persuasive evidence that all organisms have evolved from a common ancestor. Indeed, the evolutionary

TABLE 12.1 The genetic code as it applies to mRNA

UUU	Phe	UCU	Ser	UAU	Tyr	UGU	Cys
UUC	Phe	UCC	Ser	UAC	Tyr	UGC	Cys
UUA	Leu	UCA	Ser	UAA	Stop	UGA	Stop
UUG	Leu	UCG	Ser	UAG	Stop	UGG	Trp
CUU	Leu	CCU	Pro	CAU	His	CGU	Arg
CUC	Leu	CCC	Pro	CAC	His	CGC	Arg
CUA	Leu	CCA	Pro	CAA	Gln	CGA	Arg
CUG	Leu	CCG	Pro	CAG	Gln	CGG	Arg
AUU	Ile	ACU	Thr	AAU	Asn	AGU	Ser
AUC	Ile	ACC	Thr	AAC	Asn	AGC	Ser
AUA	Ile	AC	Thr	AAA	Lys	AGA	Arg
AUG	Met	ACG	Thr	AAG	Lys	AGG	Arg
GUU	Val	GCU	Ala	GAU	Asp	GGU	Gly
GUC	Val	GCC	Ala	GAC	Asp	GGC	Gly
GUA	Val	GCA	Ala	GAA	Glu	GGA	Gly
GUG	Val	GCG	Ala	GAG	Glu	GGG	Gly

The abbreviations of the 20 amino acids are: Ala, alanine; Arg, arginine; Asn, asparagine; Asp, aspartic acid; Cys, cysteine; Gln, glutamine; Glu, glutamate; Gly, glycine; His, histidine; Ile, isoleucine; Leu, leucine; Lys, lysine; Met, methionine; Phe, phenylalanine; Pro, proline; Ser, serine; Thr, threonine; Trp, tryptophan; Tyr, tyrosine; Val, valine.

relatedness of organisms is further strengthened by the universal nature of the whole genetic mechanism – the structure of DNA, its mode of replication, and the mechanisms of information flow from gene to protein being described in this chapter.

BOX 12.2 EXTRANUCLEAR INHERITANCE

Both mitochondria and chloroplasts contain their own genome. It codes for some of the proteins required in these organelles (respiratory proteins in the mitochondria and photosynthetic enzymes in the chloroplasts), along with tRNAs, rRNAs, enzymes, and proteins involved in translation of these proteins, as mitochondria and chloroplasts have their own translation machinery. Indeed, the general features of these organelles' transcription and translation components are more similar to those in bacterial cells than the eukaryotic cytoplasm in which they are found. Such observations lend support to the **endosymbiotic theory** of the origins of these organelles (i.e. that they were once free-living bacteria that invaded and subsequently established mutually beneficial relationships within proto-eukaryotic cells).

In keeping with a prokaryotic origin for these organelles, their DNA molecules are generally circular and not aggregated with histone or similar proteins. Multiple copies are present (e.g. each human mitochondrion has, on average, 10 identical DNA molecules). This means that metabolically active cells, with large numbers of mitochondria, possess thousands of copies of the mitochondrial genome. The genomes of these organelles are also much smaller than their nuclear counterparts; for this reason, they were among the first DNA molecules to be sequenced. The human mitochondrial genome contains 16,569 base pairs compared to the estimated 3 billion of the nuclear genome!

The discovery in the 1950s that DNA was present in mitochondria and chloroplasts provided an explanation for some unusual, non-Mendelian, inheritance patterns that had been puzzling geneticists for half a century. Occasionally, traits were observed that seemed to be inherited through the female line; for example in 1909, Carl Correns (1864–1933) was investigating the inheritance of leaf color in the four o'clock plant, *Mirabilis jalapa*. He was perplexed by the fact that the progeny phenotype was always the same as the female plant. Other examples of traits, seemingly inherited through the female line, continued to be discovered. Now we know that leaf color in this plant is controlled by a chloroplast gene. Chloroplasts and mitochondria are **maternally inherited**, because it is the egg that contributes the cytoplasm to the zygote and therefore for the next generation. Any traits controlled by genes within either the mitochondrial or chloroplast genome must, therefore, be inherited from the female parent. A few rare human conditions are inherited in this way. Leber hereditary optic neuropathy (LHON) is the result of a mutation in the mitochondrial NADH dehydrogenase gene. LHON is characterized by rapid loss of vision as the result of optic nerve death. It typically begins in the third decade of life and is usually irreversible. The pedigree below illustrates the maternal inheritance of this mitochondrial disease within a family.

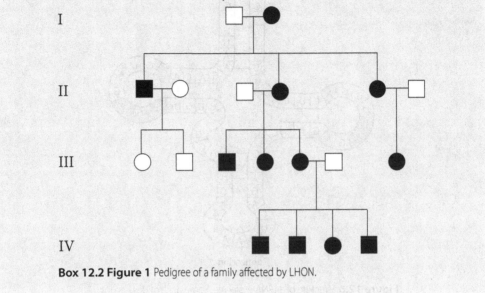

Box 12.2 Figure 1 Pedigree of a family affected by LHON.

12.7   Producing the protein: translation

During translation, the nucleotide sequence of an mRNA molecule is converted into the amino acid sequence of a polypeptide chain. It is a complex process involving the three main types of RNA – mRNA, tRNA, and rRNA (Section 12.4), as well as many proteins. It occurs in the cytoplasm at ribosomes. Ribosomes consist of rRNA and proteins, organized into a smaller and larger subunit. Some of the proteins function as structural components of the ribosomes. Most have roles to play within translation. rRNA itself is of four main types (5, 5.8, 18, and 28S rRNA) in eukaryotic ribosomes and three types (5, 16, and 23S) in prokaryotic ribosomes. (The "S" in 28S and 23S is the sedimentation coefficient, which indicates the rate of sedimentation in a centrifuge in relation to molecular weight and three-dimensional shape.) During translation the ribosome binds to, and correctly orientates, the mRNA and tRNA molecules for accurate and efficient translation.

Although the mRNA provides the instructions for synthesis of a polypeptide, it cannot bind directly to amino acids. Instead, mRNA interacts with tRNA molecules that carry amino acids and brings them to the mRNA as they are required during translation of the encoded instructions. Each tRNA molecule bonds with and carries one amino acid. Any one tRNA molecule is always loaded with the

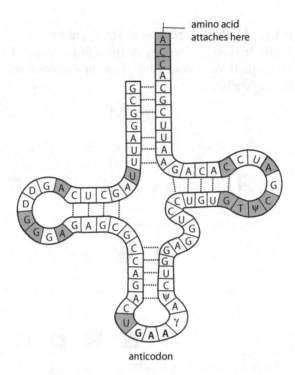

Figure 12.6 Structure of a tRNA molecule. Some nucleotides occupy the same positions in all tRNAs; these are shown in grey. Others vary according to the particular tRNA. The symbols T, D, γ and ψ represent unusual nucleotides found in tRNAs.

same type of amino acid. The type is determined by a tRNA's **anticodon** – a specific triplet of nucleotides within the molecule (Figure 12.6).

A tRNA molecule generally consists of a sequence of between 80 and 90 ribonucleotides. Segments of the single-stranded transcript are complementary. Thus, the relevant bases pair and so initiate folding of the molecule into a characteristic cloverleaf shape within which there are both double-stranded and single-stranded regions, and two key functional areas – an attachment site for an amino acid and the anticodon (Figure 12.6). In addition to determining the amino acid that a tRNA molecule carries, the anticodon is also the site of interaction between a tRNA and an mRNA molecule during translation.

Figure 12.7 depicts the process of translation. It starts when the small subunit of a ribosome binds to the initiation site, codon AUG, near the 5′ end of an mRNA molecule. This binding exposes two mRNA codons to the large ribosome subunit, within which there is room to accommodate a pair of tRNA molecules. During translation the tRNA carrying the peptide being synthesized occupies the **P site**, and the incoming tRNA with its amino acid binds to the **A site**. Initially, two tRNAs, whose anticodons are complementary to the exposed mRNA codons, enter the ribosome and hydrogen bond with the mRNA codons. In the example depicted in Figure 12.7, the two tRNAs must have anticodons UAC and CGG. They will be carrying the amino acids methionine and alanine, respectively. The first peptide bond can now be formed. Using the enzyme peptidyl transferase, methionine, the amino acid from the tRNA in the P site, is transferred and bonded to the amino acid attached to the second tRNA, in the A site.

Following formation of a peptide bond, the tRNA that lost its amino acid is released into the cytoplasm, where it will quickly pick up a replacement. The

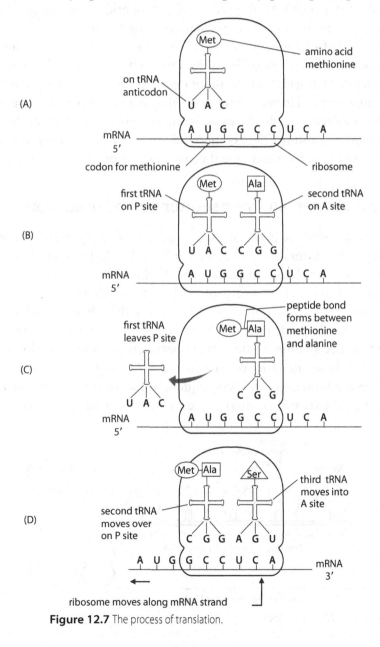

Figure 12.7 The process of translation.

ribosome then moves three bases along the mRNA, so that the next codon is within the A site and is exposed for translation (Figure 12.7c). A third tRNA whose anticodon is complementary to the next mRNA codon now enters the ribosome. A second peptide bond forms, with the growing peptide chain being transferred and linking to the amino acid on the more recently entered tRNA. A peptide of three amino acids has now been produced.

The cycle of tRNA binding, peptide formation, and tRNA release continues until a stop codon is reached. This signals to the translation machinery that its job is complete. The mRNA, ribosome, and polypeptide separate, and the polypeptide assumes its functional conformation. This folding may occur unaided or with the help of **chaperone** proteins. Post-translational modification of the newly synthesized polypeptide also occurs frequently, such as cleavage of short sequences and phosphorylation and other chemical modifications of amino acids. Generally, during translation, a number of ribosomes bind simultaneously to the same mRNA molecule, each ribosome being at a different stage in the synthesis of the encoded protein. This compound structure is known as a **polysome** (Figure 12.8) and helps to maximize the amount of a polypeptide synthesized.

12.8   The causes and consequences of mutation

This chapter has described how the nucleotide sequence of genes determines the structure of polypeptides via the intermediary of mRNA. The flow of information from DNA to protein therefore needs to be highly efficient. The consequences of any failures in this system are profound. Just one changed amino acid in a protein can prevent its functioning. This is because the efficient working of most protein molecules depends upon their precise three-dimensional forms, which result from a range of subtle interactions between different constituent amino acids (Box 12.1). One different amino acid in a protein molecule can change the nature of these interactions and, consequently, the resulting form of the protein. Thus, it is of paramount importance that the translation machinery works

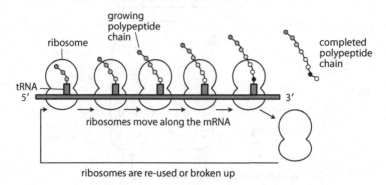

Figure 12.8 A polysome.

efficiently and, indeed, that the information the ribosomes receive is accurate. Transcription must be error free and, perhaps most critical of all, the stored information, represented by the nucleotide sequence of DNA, must remain complete and uncorrupted.

So important is the accurate maintenance of the stored information that a range of surveillance mechanisms have evolved in living organisms to monitor and correct any changes to the nucleotide sequence. As discussed in Section 11.6, most DNA polymerases also proofread newly synthesized DNA to check for any incorrectly inserted nucleotides during DNA replication. If detected, such nucleotides are excised and replaced by the correct nucleotides. Other enzymes survey DNA during normal cell functioning. Spontaneous *in vivo* modification to base structure occurs regularly. Each day, many foreign chemicals, with the potential to damage DNA, enter the bodies of living organisms. Similarly, ultraviolet (UV) and other radiation penetrates into cells and can change the nucleotide sequence. The cells, however, are ready for these challenges to the integrity of the hereditary information. Several different sets of enzymes are present in cells to detect and repair structural alterations to DNA (Table 12.2). Box 12.3 shows

TABLE 12.2 The four main mechanisms of DNA repair

REPAIR SYSTEM	TYPE OF DAMAGE REPAIRED	DETAILS
Base excision	Abnormal or modified bases	• DNA glycosylase recognizes and removes altered base • An endonuclease nicks and removes a section of nucleotides • The exposed gap is infilled by a DNA polymerase and sealed by DNA ligase
Nucleotide excision	Pyrimidine dimers or other physical DNA distortions	See Box 12.3
Mismatch repair	Misincorporated nucleotides that escaped proofreading during DNA replication	• An enzyme system binds to the mismatch and removes the misincorporated base • In prokaryotes, the enzyme has been shown to bind to the newly synthesized, unmethylated strand • A DNA polymerase resynthesizes correct DNA
Recombination	Double-stranded DNA breaks produced by ionizing radiation	• Only occurs during S and G_2 phases of the cell cycle • The cut single strands invade the newly synthesized undamaged sister chromatid whose DNA is a template for the synthesis of new correct DNA

BOX 12.3 THE REPAIR OF UV-DAMAGED DNA

Exposure to UV radiation causes the formation of **pyrimidine dimers** (covalently bonded thymine or cytosine bases) that deform DNA. The distortion is detected by complexes of enzymes scanning DNA for any structural irregularities. They initiate removal of the dimers and a few adjacent nucleotides by endonucleases, and their replacement by a repair DNA polymerase, using the complementary DNA strand as a template.

We all form potentially dangerous pyrimidine dimers in the DNA of our skin cells. Fortunately these dimers are normally repaired, unless we suffer from **xeroderma pigmentosum**. This disease is produced by a fault in one of the nucleotide excision repair enzyme complex. It is characterized by dry and scaly skin (xeroderma), and by extensive freckling and abnormal skin pigmentation (pigmentosum). The risk of skin tumors is increased 1000-fold. Other diseases result from deficient DNA repair, such as Cockayne and Fanconi syndromes and ataxia telangiectasia.

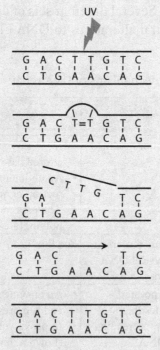

Box 12.3 Figure 1 Repair of pyrimidine dimers.

details of the repair of UV-damaged DNA and illustrates the principle by which repair systems generally work:

- Recognition enzymes detect the DNA damage.

- Nucleases excise the damaged nucleotides.

- DNA polymerases infill the gap with the correct nucleotides (determined from the complementary strand).

- DNA ligase reseals the gap.

No system, however, is totally error free. Changed nucleotides do, occasionally, escape detection, with the result that faulty mRNAs are sent to the ribosomes; even so, the occasional faulty protein is not too serious. It will be outcompeted by many more fully functional proteins, and is generally rapidly recognized by proteases and degraded. Major problems occur, however, if every copy of a particular protein is faulty. This can occur if the change or **mutation** occurred within a germ-line cell – the cell type that produces gametes. In such cases the fault is present in the DNA of all cells of the next and subsequent generations. Every copy of the encoded polypeptide is changed, often with devastating consequences.

One of the first inherited human diseases discovered to be the result of a single changed amino acid was sickle cell disease. In 1956, it was shown that an amino acid exchange had occurred in the β-globulin component of the oxygen-carrying pigment, hemoglobin. The sixth amino acid, glutamic acid, in a chain of 146, is replaced by valine. The consequences of such a seemingly minor change are profound. The resultant hemoglobin molecule is considerably less efficient at transporting oxygen around the human body and so individuals suffer from severe anemia. At low oxygen levels the changed character of the hemoglobin molecule pulls the normally biconcave red blood cells into a sickle shape (Figure 12.9A). Sickled red blood cells can no longer flow easily through capillaries, which then become blocked. Tissues become starved of oxygen and a painful "sickle crisis" results, which if untreated can be fatal. At a minimum it can cause permanent damage to vital organs. All this because one amino acid is changed in a protein!

The discovery of the biochemical basis of sickle cell disease occurred at the time scientists were unraveling the secrets of the genetic code. Soon the change to the

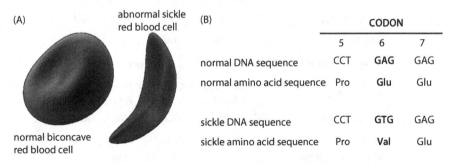

Figure 12.9 Sickle cell anemia. (A) Changed shape of red blood cells. (B) Altered DNA and amino acid sequence.

β-globulin molecule was understood at the DNA level. It was the consequence of a change to the nucleotide sequence within the globulin gene. An uncorrected **nucleotide substitution** had occurred within the codon representing the sixth amino acid, glutamic acid, in the β-globulin chain as shown in Figure 12.9, so that it now coded for a different amino acid, valine. The amino acid sequence of a protein is not, however, always changed by a nucleotide substitution. This is because most amino acids are represented by more than one codon (Table 12.1). If the change in the β-globulin gene had been from GAG to GAA (instead of GAG to GTG), there would have been no change of amino acid, as both GAG and GAA encode glutamic acid. In contrast there are a few substitution mutations (known as **nonsense** mutations) that have a more major effect as they change an amino acid encoding codon to a stop codon. Translation of the encoded polypeptide is prematurely halted!

Many diseases and other conditions in living organisms are now known to be related to nucleotide substitutions. A single alteration in the amino acid sequence of a polypeptide generally has serious consequences for the individual, as has been illustrated by the fate of individuals with sickle cell disease. Not surprisingly, cells, and therefore individuals, rarely cope with the multiple amino acid changes that occur within a polypeptide chain, if a single nucleotide is **deleted** or **inserted** into the stored information for a particular molecule. The consequence of a deletion or of an insertion is to change all codons, and therefore all amino acids, subsequent to the mutation site. This is because the **reading frame** of the message has shifted; thus, we also refer to these classes of mutation as **frameshift mutations**. During translation the message is decoded in threes. For every adjacent three nucleotides, one amino acid is added to the growing polypeptide chain. Ribosomes do not have any mechanisms to assess the validity of the message they receive. They cannot know if a message has been corrupted by the addition or removal of a nucleotide (Figure 12.10). The effective result of a frameshift mutation is a different polypeptide, which cannot fulfill its intended function.

normal DNA	A T C T G G C A C T A T G G A T A G A C C G T G A T A C C T				
normal mRNA	AUC	UGG	CAC	UAU	GGA
polypeptide	Ile	Trp	His	Tyr	Gly
frameshift mutation			A T inserted		
mutant DNA	A T C A G G C A C T A T G G A T A G T C C G T G A T A C C T				
mutant mRNA	AUC	AUG	GCA	CUA	UGGA
polypeptide	Ile	Met	Arg	Leu	Trp

Figure 12.10 The consequences of a frameshift mutation. Note the change of amino acid sequence following the mutation.

Ultimately, we are interested in understanding the link between genotype and phenotype – how the stored information leads to the expression of different phenotypes and, indeed, of a whole organism. Towards the end of the 1960s, limitations in available technology started to become apparent. The genetic code had been elucidated and the basic features of transcription and translation worked out. Molecular geneticists wanted to examine genes in more detail, but were becoming increasingly frustrated by the lack of suitable techniques. The development in the early 1970s of a whole new methodology, known as **recombinant DNA technology**, retrieved the situation. Indeed it also spurred new, previously unimagined, commercial possibilities. Chapter 13 outlines the principles of recombinant technology and explores some of its applications, along with other new technologies that have developed in recent decades.

Summary

- There is an information flow from genes to proteins.

- The information is contained within the linear nucleotide sequence of a gene. Every three adjacent nucleotides, a codon, specifies an amino acid.

- The information is carried by a mobile intermediary, messenger RNA (mRNA), to the ribosomes where proteins are synthesized.

- mRNA is one of three main types of RNA; there are also transfer RNA (tRNA) and ribosomal RNA (rRNA).

- rRNA is a key component of a ribosome, where assembly of amino acids into a polypeptide chain occurs. tRNA molecules bring amino acids to the ribosomes as they are required during translation.

- All RNA molecules consist of a single polynucleotide chain, produced by the process of transcription. One separated DNA strand acts as a template for synthesis of a complementary copy.

- During translation, there is a progressive reading of the nucleotide sequence of mRNA – one amino acid being added at a time to a growing polypeptide chain.

- The nucleotide sequence of a gene is constantly surveyed by cellular enzymes for any changes to the stored information. Detected changes are corrected by repair enzymes.

- Undetected changes represent mutation, which are of two main types – nucleotide substitution and frameshift.

- A mutation can alter the amino acid sequence of a protein and thus prevent its functioning.

Problems

1. Where are DNA and RNA found in a eukaryotic cell? Distinguish chemically, structurally, and functionally between the two molecules.

2. Name the molecules that:

   (a) Display an anticodon.

   (b) Are synthesized by RNA polymerases.

   (c) Have a cloverleaf structure.

   (d) Possess the genetic information during protein synthesis.

   (e) Contain exons and introns.

   (f) Get charged with an amino acid.

3. 5'-GGAACCCAG-3' is the sequence of bases of a short length of DNA. Reading the sequence from left to right, give:

   (a) The base sequence that will be produced as a result of transcription of this piece of DNA.

   (b) The three bases of the tRNA that will correspond to the underlined bases.

4. Why do mutations involving the deletion of a base usually have greater effects than those involving substitution of one base for another?

5. Identify the stages during transcription and translation that involve complementary base pairing.

6. Compare and contrast the structure and roles of tRNA and mRNA during translation.

7. Arrange the following terms according to their hierarchical relationship to each other: chromosomes, genomes, nucleotides, genes, codons, exons.

8. The H1 histone protein of the African clawed frog, *Xenopus laevis*, is composed of 193 amino acids. What is the minimum number of nucleotides in an mRNA molecule coding for this polypeptide?

9. Match each term in the right-hand column of the table to its definition in the left-hand column.

Definition	Term
1. A group of three mRNA bases that specifies an amino acid when translated	A. Transcription
2. Most amino acids represented by several codons	B. Translation
3. Removal of introns from primary transcript	C. Codon
4. UAG, UAA, or UGA	D. Collinearity

Definition	Term
5. Insertion or deletion of a number of nucleotide pairs, other than a multiple of three	E. Reading frame
6. The process during which an RNA molecule is synthesized from a DNA template	F. Intron
7. The reading of successive mRNA base triplets as codons	G. Splicing
8. The linear correspondence between the order of amino acids in a polypeptide chain and the linear sequence of nucleotides in the encoding gene	H. Frameshift mutation
9. A nucleotide sequence that is excised from the primary mRNA transcript	I. Stop codon
10. The process during which an amino acid sequence is assembled according to the information specified by a mature mRNA	J. Degeneracy of the genetic code

10. Use Figure 12.1 to help you complete the following table:

		G										
					A	C	T					**DNA**
A	G						U					**mRNA transcribed**
								C	U	U		**tRNA anticodon**
		Tryptophan										**Encoded amino acid**

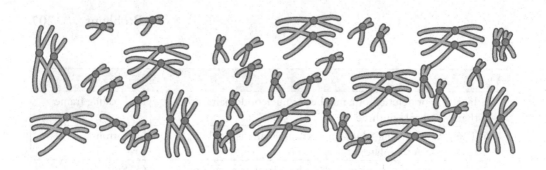

Manipulating DNA

There are few things more upsetting than witnessing the pain and distress of a child suffering a hemophiliac crisis. For centuries parents have had to look on helplessly while their child bleeds, until eventually the blood begins to clot. Now, thanks to modern DNA technology, this need never happen. We now have the molecular toolkit available to make a functional copy of the relevant gene from the genome of an individual whose blood clots properly and insert it into plasmid DNA, which is then introduced into bacterial cells. These recombinant bacteria cells can then be cultured on a large scale and programmed to produce large quantities of the vital blood-clotting protein. Thus, a hemophiliac can have an emergency supply of the critical clotting factor constantly within easy reach and need never suffer such distress. Gene therapy for hemophiliacs (i.e. the attractive prospect of permanently establishing a normal clotting factor gene in the tissues of a hemophiliac) is also being actively researched.

The ability to cut-and-paste genes at will into the genomes of different organisms is an illustration of recombinant DNA technology ("genetic engineering").

This chapter introduces:

• The key methods involved in manipulating genes

- The range of experimental and commercial uses of recombinant DNA technology
- Polymerase chain reaction (PCR) and DNA sequencing

13.1   An overview of recombinant DNA technology

Recombinant DNA technology refers to any process that modifies an organism's genotype in a directed and predetermined way. Whether we are producing a recombinant bacterium capable of producing human growth hormone or a soybean plant resistant to the herbicide glyphosate, the approach is essentially the same and can be summarized in the following stages:

1. DNA containing the gene(s) in which we are interested is purified. This DNA is referred to as the **donor DNA**.

2. The donor DNA is chopped into approximately gene-length fragments using **restriction enzymes**.

3. The desired donor fragment is inserted into a self-replicating DNA molecule, which serves as a **vector** or carrier molecule. The vector with its inserted DNA fragment is referred to as the **recombinant DNA molecule**.

4. The vector is introduced into a host cell where either the vector is maintained as a self-replicating unit or the donor DNA enters the host cell genome.

5. As the host cell replicates, the recombinant DNA molecules are passed to progeny cells. The result is a molecular clone of the inserted DNA fragment.

6. The final outcome depends upon the underlying purpose of the manipulation. The cloned DNA may be recovered from the host cells, purified, and analyzed, as, for example, in many genome sequencing projects. Alternatively, the encoded protein may be expressed, purified, and sold commercially. Harvesting clotting Factor VIII for hemophiliacs or insulin for diabetics are two examples of this use of genetic engineering. If the aim of the manipulation was to confer a permanent new phenotype on the host organism, as with maize made resistant to insect attack, then the recombinant DNA acts *in situ*.

The different stages outlined above will now be considered in more detail.

13.2   Restriction enzymes

Restriction enzymes are a critical component of the DNA cloning toolkit. These molecular "scissors" are produced by bacteria; their natural role is to prevent or "restrict" bacteriophage (viruses that infect bacteria) attack by chopping up the infecting DNA. The discovery of restriction enzymes in the late 1960s by Werner Arber, Daniel Nathans, and Hamilton Smith made molecular cloning a possibility, and earned them the 1978 Nobel Prize for Physiology or Medicine.

Around 4000 different restriction enzymes have now been isolated from a wide range of bacterial species; several hundred enzymes are available commercially. Restriction enzymes are endonucleases (i.e. they cleave DNA internally). They recognize and cut at specific nucleotide sequences, generally four to six nucleotides in length. Typically, a restriction site is **palindromic** – the recognition sequence is repeated nearby in reverse complementary orientation on the opposite DNA strand. For example, the palindromic recognition site of the restriction enzymes *Hin*dIII from the bacteria *Haemophilus influenzae* is:

5′-AAGCTT

3′-TTCGAA

Restriction endonucleases either cut cleanly across both strands producing **blunt ends** or make a staggered cut, generating single strands, or **sticky ends** (Table 13.1).

Generally, staggered-cutting restriction enzymes are used to produce recombinant DNA molecules. Owing to DNA's complementarity, these sticky ends have a natural affinity for each other or similar single strand ends. This property, combined with the fact that restriction enzymes recognize their restriction sequence whatever the source of the DNA, is the key to the use of these enzymes in recombinant DNA technology. Both the source DNA and vector DNA is cut with the same staggered-cutting restriction enzyme. The resulting single-stranded overhangs enable hybridization of the DNA fragment to the cloning vector's DNA (Section 13.4).

TABLE 13.1 Recognition sites of the two restriction enzymes *Eco*RI and *Hae*III from the bacteria *Escherichia coli* and *Haemophilus aegyptius* (an enzyme's name is based on the bacterial species name and order of discovery)

ENZYME	BACTERIAL SPECIES	RECOGNITION SEQUENCE	CUT
*Eco*RI	*E. coli*	5′-GAATTC	5′-G AATTC-3′
		3′-CTTAAG	3′-CTTAA G-5′
*Hae*III	*H. aegyptius*	5′-GGCC	5′-GG CC-3′
		3′-CCGG	3′-CC GG-5′

13.3   Vectors

Restriction enzymes enable genomic DNA to be chopped into pieces. Once the desired DNA fragment has been identified, a carrier DNA molecule or **vector** is employed to introduce it into the relevant host cell. Over the last 30 years a range of vectors has been developed. The most common vector is a **plasmid**.

Plasmids are naturally occurring, small, double-stranded circular DNA molecules within the cytoplasm of bacterial and a few eukaryotic species (Figure 13.1). They replicate independently of the bacterial chromosome; the number of copies of a plasmid varies, depending upon plasmid type and circumstances. Plasmids carry genes that aid the survival of bacteria under adverse conditions. Their most important role (for humans and animals) is carrying antibiotic resistance genes. They also may possess genes that enable bacteria to degrade pollutant compounds, produce antibacterial proteins called colicins, and confer increased pathogenicity. The antibiotic resistance and autonomous replication of plasmids are the two characteristics that have been exploited in the development of DNA recombinant technology.

The desired donor DNA fragment is inserted into a plasmid at a restriction enzyme site. The plasmid is then introduced into the host bacterial cell. As the plasmid replicates, multiple copies of the donor DNA are produced. Antibiotic resistance represents a **marker gene**, indicating the presence of a plasmid (with a cloned gene) in a bacterial cell. Figure 13.1 shows the key features of a cloning plasmid (indeed, of any cloning vector):

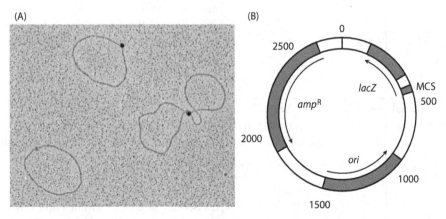

Figure 13.1 Features of a cloning plasmid. (A) Scanning electron micrograph of plasmids extracted from *E. coli*. From Inman, R. B. and Jackson, J. F. (1989) *Gene* **84**, 221–226. With permission from Elsevier. (B) Details of a common cloning plasmid: pUC19 (2686 bp): note the origin of replication, *ori*, together with the antibiotic (ampicillin)-resistance gene (*amp*^{R}), *lacZ*, and multiple restriction sites (MCS). The *lacZ* gene encodes β-galactosidase – part of a screening strategy to detect plasmids with cloned DNA.

- A selectable marker, generally antibiotic resistance
- An origin of replication
- Restriction enzyme sites

The choice of cloning vector depends upon several factors; in particular, upon the size of the DNA that is being cloned and its destination. Bacterial plasmids accept relatively small fragments, just 5–10 kb in size. A human gene can be 1–2 Mb in size (1 kb = 10^3 bases and 1 Mb = 10^6 bases). Thus, over the years, a range of different cloning vectors has been developed. Table 13.2 gives a selection, including some key eukaryotic vectors.

Bacteriophages ("phages") are the viruses of bacteria. Their value in cloning is that they have the genetic apparatus to infect and multiply within a bacteria cell. Commonly, lambda (λ) phage or a hybrid version – a **cosmid** – is used as a cloning vector. Cosmids and **bacterial artificial chromosomes (BACs)** have proved

TABLE 13.2 Main cloning vectors

VECTOR	SIZE OF INSERTED DNA	HOST CELL TYPE	VECTOR DETAILS
Plasmid	5–10 kb	Bacteria	Self-replicating double-stranded DNA molecules in bacterial cytoplasm
Lambda (λ)	15–20 kb	Bacteria	A bacteriophage
Cosmid	50 kb	Bacteria	A hybrid vector consisting of approximately one-third of the λ genome (i.e. genes that control entry into bacterial cells) and plasmid genes for replication; loss of most of the λ genes leaves room for bigger DNA inserts
Bacterial artificial chromosome (BAC)	300 kb	Bacteria	An artificially created vector based on the F plasmid
Yeast artificial chromosome (YAC)	1–2 Mb	Yeast and mammalian cells	A hybrid chromosome containing a centromere, a telomere at each end, an origin of replication, a cluster of restriction (cloning) sites, and marker genes
Retrovirus	8 kb	Mammalian cell	Its single-stranded RNA genome, with a transcript of donor DNA, is reverse transcribed into DNA within a host cell where it integrates into the host cell chromosome
Ti plasmid	180 kb	Crop plants	Plasmid of soil bacterium; a segment, T-DNA, readily inserts into host chromosomes carrying donor DNA

useful cloning vectors for the human and other genome projects. DNA recombinant techniques were developed using prokaryotic systems. However, if we are interested in the expression of eukaryotic genes, a eukaryotic cloning organism is more desirable. Brewers' yeast, *Saccharomyces cerevisiae*, is an ideal eukaryotic host species. It is unicellular, and so can be grown and manipulated in a similar way to bacterial cells. Much is known of its genetic system and it contains a plasmid – the 2-μm circle. A linear **yeast artificial chromosome** (**YAC**) has also been created and used widely as a cloning vector.

13.4   Creating a recombinant DNA molecule

A cloning plasmid is chosen that contains the nucleotide sequence recognized by the restriction enzyme that was used to generate the donor DNA fragments. Alternatively, relevant recognition sequences can now be engineered into a plasmid (Figure 13.1). The chosen plasmid is treated with the restriction enzyme. This opens up (**linearizes**) the plasmid and produces sticky ends. These sticky ends will be the same as those of the donor DNA. When the donor fragments and linearized plasmid are mixed, complementary base pairing occurs between the single-strand overhangs produced by the restriction enzyme. DNA ligase is added to seal the joints. The process is summarized in Figure 13.2.

13.5   Introducing the recombinant DNA molecule into a host cell

Once the recombinant DNA molecules have been produced, they are introduced into host cells. Various techniques exist. The simplest is the treatment of bacteria cells with calcium chloride to facilitate the uptake of plasmids. Alternatively, a process called **electroporation** can be used, when cells are stimulated with a brief, weak electric shock. This causes the cell membrane to become temporarily permeable to DNA. Bacterial cells that have taken up recombinant DNA are referred to as **transformed cells**. Transformed cells have to be identified and separated from those cells that lack a plasmid. A common way to distinguish between the two types of cells makes use of the antibiotic resistance conferred on a host cell by the presence of a plasmid.

Bacteria can be grown in liquid culture or on solid nutrient plates where they form **colonies**. All bacterial cells within a colony are derived from a single ancestral cell; thus, all cells of a colony, and the plasmids they contain, are genetically identical. To screen for those cells that have taken up a plasmid, the bacteria can be plated on antibiotic-containing agar. Only those bacteria with a plasmid, and so a resistance gene, will grow. When cloning, not all plasmids will be recombinant plasmids. Thus, an extra step (the "blue/white test") is needed to identify

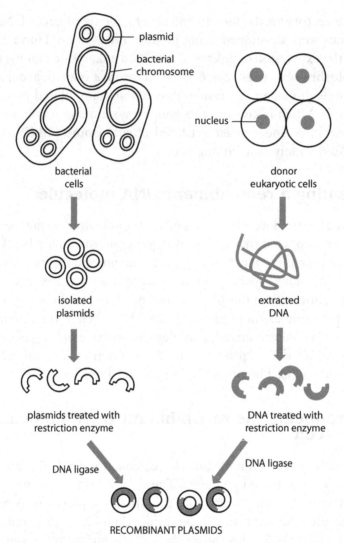

Figure 13.2 Construction of recombinant plasmids.

those bacteria with recombinant plasmids. The organic compound X-gal distinguishes bacteria with recombinant or non-recombinant plasmids (Figure 13.3).

X-gal produces galactose and a blue product when cleaved by the enzyme β-galactosidase. The *Escherichia coli* gene *lacZ* encodes β-galactosidase. Many cloning plasmids have the *lacZ* gene transferred into their genome with restriction sites engineered into it. Therefore, in recombinant plasmids, *lacZ* is disrupted by the inserted DNA fragment. Thus when bacteria with recombinant plasmids are plated on media containing X-gal, there is no β-galactosidase produced and so the bacteria are white. Bacteria with non-recombinant plasmids have a functional enzyme and so are blue. The β-galactosidase acts as a **reporter gene**, indicating whether or not a bacterial colony contains recombinant plasmids.

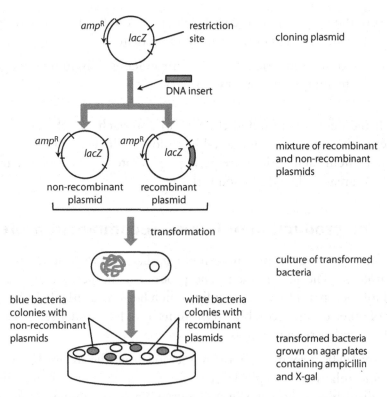

Figure 13.3 Identification of bacteria with recombinant plasmids by blue/white X-gal screening.

The use made of the transformed cells with their recombinant DNA varies according to the host organism and, of course, the reasons for their production. In 1973, Herbert Boyer and Stanley Cohen performed the first successful recombinant DNA cloning experiment; introducing into a population of *E. coli* previously resistant to a single antibiotic, plasmids with a second antibiotic resistant gene. Forty years later, the range of uses of recombinant technology is enormous. It has revolutionized all fields of experimental biology and spread far beyond the research laboratory into the commercial market, supporting a booming biotechnology industry. The following list gives a selection of the range of uses of recombinant DNA technology:

- The efficient production of useful proteins in large quantities, such as clotting Factor VIII, human growth hormone, and insulin, with their obvious medical applications, or rennin for the cheese industry

- Generation of DNA sequences as research tools or for use in medical diagnosis, such as prenatal diagnosis or carrier testing

- The introduction of genes into plants and animals to produce new, desired characteristics, such as β-carotene synthesis genes into the genome of rice as part of the programme to counter vitamin A deficiency or herbicide resistance genes into the genomes of many crop plants

- The correction of genetic defects by introducing fully functional genes into cells – the technique of gene therapy

Although the details may differ, the overall approach to achieving any of the outcomes in the above list is remarkably uniform. The use of recombinant techniques in medicine and agriculture perhaps generates the fiercest debate; aspects of these two areas will be discussed further.

13.6   The production of human recombinant proteins

Initially, biotechnology companies employed bacteria to produce their recombinant proteins. The first human gene product manufactured in this way and licensed for therapeutic use was insulin, which became available in 1982. A huge number of other useful recombinant proteins have followed, used both clinically (Table 13.3) and as research tools.

In the early 1980s it seemed almost miraculous that the means had been found of providing, relatively cheaply, large supplies of much needed medicines. Soon, however, disturbing reports began to emerge. Some individuals using recombinant insulin and growth hormone were suffering allergic reactions. It seemed that there were imperfections in the system; for example, eukaryotic proteins did not always fold properly into their correct three-dimensional form in the prokaryotic bacterial cytoplasm. Bacterial cells could not fully process and modify the recombinant protein if extra groups, such as phosphates or sugars, needed to be added post-translationally. Such discoveries suggested a new approach was needed to the production of recombinant proteins. It was argued that eukaryotic recombinant proteins should be produced by eukaryotic cells and, if possible, human proteins by mammalian cells. Attention was focused on the possibility of

TABLE 13.3 A selection of human recombinant proteins available for clinical use

GENE PRODUCT	CONDITION TREATED
Epidermal growth factor	Burns and skin grafts
Erythropoietin	Anemia
Factor VIII	Hemophilia A
Interferon-α	Chronic hepatitis C
Growth hormone	Dwarfism
Insulin	Type I diabetes
Interleukin-2	Cancer
Tissue plasminogen activator	Acute myocardial infarction

modifying the genome of livestock so that they produced much needed human proteins in their milk. One of the first proteins to be successfully produced in this way was the enzyme α_1-antitrypsin, in the milk of Tracy the sheep. Lack of α_1-antitrypsin produces a hereditary form of emphysema that is progressive and eventually fatal.

Two major problems were faced by geneticists when developing these new recombinant technologies:

- How to introduce the relevant human gene into the genome of a sheep or other mammal

- How to limit expression of the introduced gene to one tissue

It was soon realized that the best time to introduce foreign genes into the genome of a complex multicellular organism, such as a sheep, was at an early embryological stage, even into the zygote or unfertilized egg. In the case of Tracy, the α_1-antitrypsin gene was introduced into the zygote. To ensure that expression of the recombinant gene only occurred in mammary tissue, the α_1-antitrypsin gene was hitched to a special control region that is only operational in this tissue. The original bacterial recombinant technology still had its uses. Bacteria were used to provide multiple copies of the fusion gene to be introduced into sheep zygotes. Figure 13.4 summarizes the procedure used to produce Tracy. When the transformed zygote had divided a few times the embryo was introduced into a foster mother. Once sexually mature, Tracy was mated with a normal ram. The milk she produced had high concentrations of the protein. Tracy is an example of a **transgenic animal**.

Transgenic mice have become extremely useful models of human disease. Cattle, sheep, pigs, and goats are now used routinely as animal bio-factories. Another sheep, Polly, was the first to produce human clotting Factor IX. The cloning of the now famous Dolly was viewed as the first stage in a long-term program, aimed at developing techniques to clone efficient human protein producers such as Tracy and Polly. The rationale was that once you have an efficient protein-producing animal, it is easier to make multiple copies of this animal than to start the transformation process again! However, the creation of Dolly initiated strong debate because the technology used in her creation could, theoretically, be used in attempts to clone humans. Geneticists have, though, over the last couple of decades been working towards producing "transgenic humans" as a means of curing genetic disease (i.e. there has been much work towards perfecting the technique of **gene therapy**).

Hopes were high in the late 1980s that many serious inherited disorders could be cured by replacing deficient genes with functional copies. However, the results of two decades of intensive research and the occasional clinical trial have been

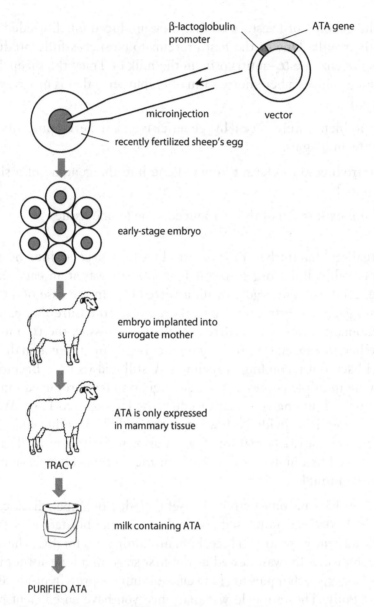

Figure 13.4 The creation of Tracy. The α_1-antitrypsin (ATA) gene was linked to the β-lactoglobulin promoter. As this promoter is only expressed in mammary glands, α_1-antitrypsin production is limited to this tissue where it is secreted into the milk, from which it is purified.

disappointing. It has proved enormously difficult to deliver successfully the modifying genes into the correct target tissue, let alone ensure the relevant gene's controlled integration into a suitable human chromosome where its long-term expression could occur. The possibility of germ-line therapy (i.e. altering all cells including those that give rise to the gametes) as the solution to this seemingly intractable problem has been considered – and rejected. The implications of modifying the genome of a germ-line cell are enormous, as all cells and so future human generations derived from of a modified germ-line cell would be changed permanently. Thus, we might achieve our goal of curing fatal genetic diseases, but at the same time we are likely to raise all sorts of other, unforeseen issues. How might society, for example, respond to the suggestion that we use germ-line gene therapy to extend intellectual potential or enhance athletic ability? The difficult ethical dilemmas created by the development of recombinant techniques are not confined to transgenic animals. Some of the most intense debate of recent years regarding the production of genetically modified organisms has focused on the development of transgenic crops.

13.7 The production of transgenic crops

Weed infestation is responsible for an estimated 10% decrease in crop productivity worldwide. The response to this problem in the recent past has been to use a vast range of different herbicides to kill these offending weeds. However, there has always been the risk of killing the crop itself. Furthermore, many herbicides are long lasting. They seep into water supplies, sometimes creating toxicity problems for humans. One of the few herbicides that is effective at low concentrations, short lived (because it is degraded by soil microorganisms), and non-toxic to humans is glyphosate. Unfortunately, crop susceptibility was too high, until recombinant DNA technology came up with a solution.

Glyphosate kills plants because it inhibits the action of a crucial chloroplast enzyme, EPSP synthetase, which is important in amino acid biosynthesis. Certain strains of *E. coli* have a changed EPSP synthetase enzyme and are resistant to the inhibitory effects of glyphosate. Why not put a copy of this bacterial gene into the genome of different crop plants? This is just what Monsanto and other biotech companies achieved in the early 1990s. An ideal system, using the plasmid present in the soil bacterium *Agrobacterium tumefaciens*, was already available for transferring genes into plant genomes.

A. tumefaciens naturally enters plants at sites of wounds, stimulating the formation of calluses. *A. tumefaciens*, like many bacteria, possesses a plasmid. It is this **Ti plasmid** that is responsible for the development of a callus, or to be more precise, one region of the plasmid genome – its **transfer DNA (T-DNA)**. When plants are infected with *A. tumefaciens* the T-DNA inserts into one of the plant's chromosomes and directs the synthesis of cell division proteins. The recombinant DNA strategy is to hitch the donor gene(s) to the T-DNA, which integrates them into the plant genome (Figure 13.5).

Cells transformed with the Ti plasmid need to be identified. For this reason the original modification usually occurs in culture. For example, *A. tumefaciens*

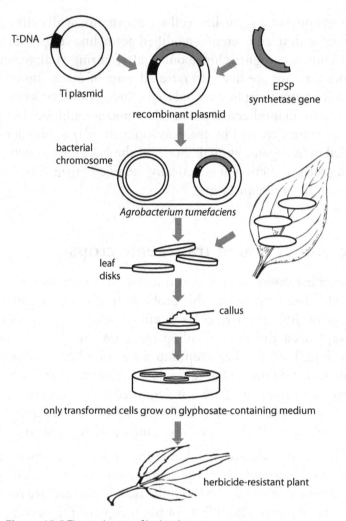

Figure 13.5 The production of herbicide-resistant crops.

containing the EPSP resistance gene within its Ti plasmid infects plant disks (Figure 13.5). When a callus appears the cells are then cultured on medium containing glyphosate. Any cells that grow must have the resistant gene. Techniques are available to nurture these undifferentiated cells to produce root and stem systems. In addition to the production of transgenic maize, soybean, and wheat resistant to the herbicide glyphosate, various other commercially useful genes have been introduced into a range of different crop plants (Table 13.4).

The scientific success in producing these **genetically modified (GM)** crops has not been matched by commercial success. Concerns soon started to be expressed over the production and uses of these crops; over the safety of foods derived from GM crops and their environmental impact. As a result, no GM crops are currently grown on a commercial scale in the UK. In 2012, just 28 countries (20

TABLE 13.4 A selection of the range of different traits that have been introduced into plants by genetic manipulation

TRAIT	EXAMPLES OF MODIFIED PLANTS
Insect resistance	Maize, cotton, potato, tomato
Herbicide resistance	Maize, soybean, cotton, rape, sugar beet, rice, flax
Virus resistance	Squash, papaya, potato
Delayed fruit ripening	Tomato
Altered oil content	Rape, soybean
Pollen control	Maize, chicory
Hepatitis B vaccine	Tobacco

developing and eight industrialized countries) worldwide grew GM crops. The USA continues to be the lead country in terms of hectares of GM crops grown, although, significantly, Brazil, China, India, Argentina, and South Africa, which together contain 40% of the world's population, collectively grew 46% of global GM crops in 2012.

Table 13.5 contrasts some of the key concerns over the growth of GM crops with what might be regarded as the benefits. From many perspectives we can argue that the goals of GM crop technology are no different to those of conventional crop manipulation, or selective breeding, which has been practiced since the beginning of agriculture. In both cases the objective is the same – to alter the genetic makeup of a plant in a predetermined "improving" direction. By employing recombinant DNA techniques this modification can be achieved much more quickly and we can widen our notion of what is a desirable gene, because genes can be introduced from any other species, be it virus, bacterium, animal, or plant. The consequences of crossing the species barrier is a new issue within the modification debate, and cannot be quickly and easily analyzed. However, frequently in the production of GM crops we are actually grappling with many issues that are equally relevant to the production of modified plants by selective plant breeding (consider the first seven risk factors in Table 13.5).

The verdict remains open on the safety and propriety of the development and use of GM crops. There seems little doubt that scientists, farmers, politicians, retailers, environmental groups, consumers, and others will continue debating the issues for many years yet. Indeed, as we develop an ever more detailed understanding of how genes work and how expression is coordinated on a cell-to-cell, day-to-day, and life-span basis, we will invent new, hitherto undreamt-of applications with an attendant spectrum of difficult ethical and social issues. Can all our knowledge also bring us closer to understanding what it means to be human? Our response will, perhaps, depend upon our personal, philosophical, or religious beliefs. Who could have predicted that Gregor Mendel's patient

TABLE 13.5 A comparison of potential benefits and concerns of growing GM crops

BENEFITS	RISKS
Targeted and reduced use of pesticides benefits pollinators and natural predators or pests	Introduced genes spreading to other crop varieties of the same species and to closely related wild species
Reduced herbicide use (e.g. one broad-acting herbicide spray instead of multiple spraying with several different herbicides)	Pathogens and pests evolving to overcome the genetically resistant crops
Better weed control reduces tillage and benefits soil biodiversity, reduces erosion, and helps moisture content	Transfer of pest-resistant traits to non-target organisms
Increased yields mean more food from a given plot, leaving more land for wildlife conservation and other purposes	Detrimental effects of herbicide resistance on plant biodiversity
The introduction of traits that enable the use of previously marginal and unproductive land	Uncertainty about how the new combinations of genes will behave
The development of foods with added nutritional value (e.g. high levels of vitamins), longer shelf-life, and better flavor and texture	Generation of plants with undesirable and unstable characteristics
The development of crops for non-food uses (e.g. valuable oils and starch – a renewable and environmentally friendly alternative)	Exposure to new, possibly harmful, allergens
The use of crops to produce edible vaccines, thus improving health care in many parts of the world	"Playing God" by the "unnatural" introduction of genes from viruses, bacteria, animals, and non-related plants into crop species

observations in the monastery garden at Brno could have led, in a little more than a century, to the rich and diverse science of today?

This chapter ends with a description of two techniques that have become the "work horses" of contemporary molecular biology:

- The polymerase chain reaction
- DNA sequencing

13.8   The polymerase chain reaction

The **polymerase chain reaction (PCR)** is a second way of producing multiple copies of a specific DNA sequence. Indeed, it has become a pivotal technique in much molecular biology. Its applications are too numerous to list! It is, for example, used to generate DNA for sequencing, in forensic DNA fingerprinting, and in the diagnosis of many human diseases Starting with a few copies of the target DNA, millions of copies can be produced in a few hours.

The basis of PCR is replication of DNA. It is catalyzed by a special DNA polymerase referred to as *Taq* **DNA polymerase**, which is derived from the thermobacterium *Thermus aquaticus* and is thermostable – retaining its catalytic properties at the high temperatures used during PCR. The other critical component of PCR is a pair of **primers** – two short single-stranded pieces of DNA (typically 18–25 nucleotides in length) – which define the target DNA region to be amplified and also provide the necessary priming DNA for the *Taq* polymerase. In addition to *Taq* DNA polymerase, a pair of DNA primers, and the target DNA, a PCR reaction mix will also typically contain all four deoxyribonucleotides and magnesium chloride, in a suitable buffer.

PCR involves three reactions, constituting the **PCR cycle**, which is repeated many times. These three reactions are (see Figure 13.6):

- **Denaturation**: the PCR reaction mixture is heated to 90–95°C. This high temperature separates the two DNA strands so that the primers can "find" their complementary (target) DNA sequences.

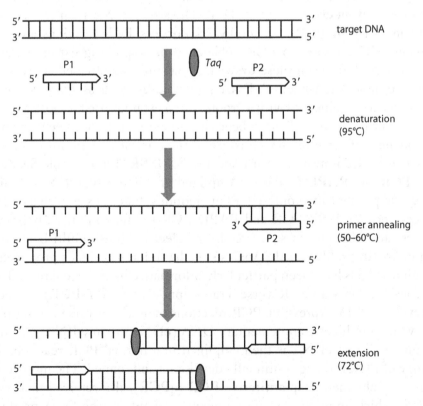

Figure 13.6 Key events of a PCR cycle. P1 and P2 are the two DNA primers; *Taq* is the *Taq* DNA polymerase.

- **Annealing**: the temperature is lowered, generally to 50–60° C, so that the primers hybridize to their target sequences.

- **Extension**: the temperature is raised to 72° C – the optimal temperature for the *Taq* DNA polymerase to synthesize a new DNA strand from the end of the annealed primer, using the original DNA as a template.

Each step in a PCR cycle generally lasts 30–60 s; the experimenter sets the time. To generate sufficient DNA for subsequent analysis, one PCR experiment typically involves 25–35 cycles; the exact number depends on factors such as initial amount of DNA and target site size. PCR cycling is automated – it occurs in a **thermocycler** in which metal blocks are programmed to rapidly change temperature. Subsequent detection and identification of PCR products is by **gel electrophoresis** (see Box 13.1).

Although PCR has proved a hugely useful molecular biology technique over the last three decades, it has a few limitations. PCR requires prior knowledge of the DNA sequence to be amplified so that primers can be designed. DNA synthesis during a PCR has a high error rate: approximately 1 in 10,000 nucleotides are misincorporated using *Taq* DNA polymerase. This is because *Taq* polymerase lacks a $3' \rightarrow 5'$ exonuclease (Section 11.6). The size of DNA fragment that can be easily amplified by PCR is generally 2 kb or less. PCR reactions are easily contaminated. This relates to PCR's efficiency in amplifying extremely small amounts of DNA! Yet even these drawbacks have been addressed successfully. For example, new "high-fidelity" DNA polymerases, such as *Pfu* DNA polymerase, have been engineered with a proofreading exonuclease for more accurate PCR. Different types of PCR have been developed that do not need any prior DNA sequence information, which is useful in situations when no or minimal genomic knowledge is available for a species. For **ISSR** (**I**nter **S**imple **S**equence **R**epeat)-**PCR** and **RAPD** (**R**andom **A**mplified of **P**olymorphic **D**NA)-**PCR**, a shorter single primer (often only 10–12 nucleotides) functions as both a forward and reverse primer. ISSR-PCR and RAPD-PCR produce multiple fragments. When these are subsequently separated by gel electrophoresis multi-band (typically three to 10) profiles result. Analysis of the different profiles produced by different individuals have been particularly informative in phylogenetic and biological conservation studies. Reverse Transcriptase PCR (**RT-PCR**) is used to detect levels of RNA expression. PCR reactions include a reverse transcriptase enzyme which produces complementary DNA (**cDNA**) to any RNA present in a cell sample. Target cDNA is then amplified in a normal PCR reaction. The technology of PCR is being continually developed and refined. The most recent major new development is **quantitative PCR** (**qPCR**), also referred to as **real-time PCR**, which amplifies and simultaneously quantifies the DNA produced from the target sequence.

BOX 13.1 PRINCIPLES OF GEL ELECTROPHORESIS

Electrophoresis separates macromolecules (i.e. DNA, RNA, or proteins) that differ in size, charge and/or conformation. Considering nucleic acids, a mixture of different sized molecules are placed in an electric field. As both DNA and RNA have a negative charge (conferred by the phosphate backbone), different molecules migrate towards the positive pole (anode) at different rates according to their different sizes – the smaller fragments moving faster. Movement of DNA or RNA occurs through a "gel" matrix. The gel is most commonly cast as a thin slab with wells at one end for loading samples. The gel is placed within an "electrophoresis tank" and covered with a buffer to maintain constant pH and which also contains ions to carry a current. For separating different DNA or RNA molecules, the gel is typically agarose or polyacrylamide.

Agarose gels are the easiest to prepare. Depending upon the agarose concentration (typically between 0.5 and 2%), DNA fragments between 100 and 20,000 bp in length can be efficiently separated. The higher the agarose concentration, the better separation of small DNA fragments. Polyacrylamide gels give better resolution of smaller DNA fragments.

Agarose gels are generally run in a horizontal "slab" format (see figure below). The sample to be analyzed is typically mixed with a loading dye, which contains something dense (e.g. glycerol) to keep the sample in the sample well, and with one or two tracking dyes, which migrate in the gel and allow visual monitoring of how far the electrophoresis has proceeded. A dye, such as ethidium bromide which binds DNA by intercalating between the DNA strands, is generally incorporated in the gel so that the DNA fragments can be located when subsequently visualized under UV light. Separated DNA molecules can be sized by running **DNA ladders** (mixtures of DNA fragments of known size) in one of the gel lanes. Size measurement and analysis are often performed with a specialized gel analysis software. DNA can be sliced out from a gel and purified for future work.

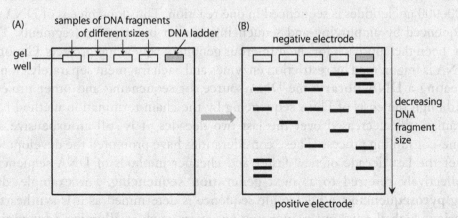

Box 13.1 Figure 1 Principles of gel electrophoresis. (A) DNA samples are loaded into gel wells. (B) After, for example, 100 V for 1 h, DNA fragments are separated relative to size.

13.9   DNA sequencing

The most common method of DNA sequencing is the **chain termination method** developed in 1977 by the twice Nobel Prize winner Frederick Sanger (hence, also known as **Sanger sequencing**). A special PCR reaction, referred to as **cycle sequencing**, produces multiple copies of *one* of the strands of the piece of DNA to be sequenced – this is achieved by only adding one primer to the PCR reaction mix. Critically, the copies of the template DNA are of different sizes. These different sized PCR products are produced because each dNTP is a mixture of the normal deoxyribonucleotides (dNTPs) and a small fraction of fluorescently labeled **dideoxyribonucleotides** (ddNTPs). ddNTPs lack the 3′-OH on the deoxyribose sugar, and when incorporated into a growing DNA chain they terminate DNA synthesis. During the extension step, *Taq* polymerase randomly adds either a dNTP or a ddNTP to the growing chain. Whenever a ddNTP is added, DNA synthesis stops. Thus, when the cycle sequencing PCR is complete, there are PCR fragments of all possible different lengths, each terminated by a fluorescing ddNTP (Figure 13.7A).

The different sized PCR fragments are then analyzed by a DNA sequencer. The collection of DNA fragments generated by cycle sequencing PCR are separated by **polyacrylamide gel electrophoresis** (**PAGE**) in a thin capillary tube. A laser then scans the gel to determine the identity of the final nucleotide of each band according to the wavelengths at which its terminating ddNTP fluoresces. The results are depicted in the form of a **chromatogram**, the characteristic DNA sequencing trace, in which each colored peak corresponds to the nucleotide at that location in the sequence: A is green, T is red, G is black, and C is blue (Figure 13.7B).

The chromatogram gives the sequence of one DNA strand; that of the opposite strand can be deduced from complementary base pairing. A DNA fragment of 500–900 nucleotides is sequenced in one reaction. Thus long pieces of DNA are sequenced by amplifying and sequencing smaller overlapping fragments. This has been the approach for the numerous genome-sequencing projects. Genomic DNA is fragmented by restriction enzymes and each fragment separately cloned, creating a **DNA library**: the DNA source for sequencing and other projects. Although the costs of DNA sequencing by the chain termination method have dramatically decreased over the last two decades, it is still an expensive and time-consuming process. These considerations have prompted the development over the last decade of new faster and cheaper methods of DNA sequencing collectively referred to as **next-generation sequencing**. For example, during **pyrosequencing**, a nucleotide sequence is determined as it is synthesized. Various high-throughput sequencing processes, such as **Illumina sequencing**, produce millions of sequences in one reaction. As well as lowering the cost of

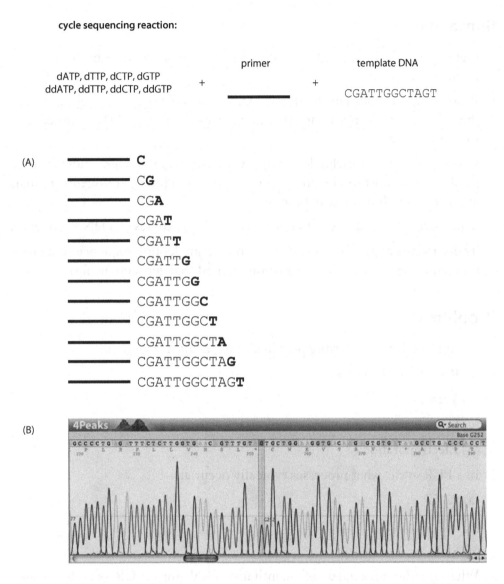

Figure 13.7 (A) The set of PCR fragments produced during cycle sequencing of a 12-nucleotide DNA fragment. Letters in bold represent the ddNTPs. Each DNA fragment begins with the PCR primer. (B) A DNA-sequencing chromatogram.

sequencing there has been concomitant development of computer software and Internet resources (collectively referred to as **bioinformatics**) to analyze the huge amount of data generated. Together, these new sequencing techniques and bioinformatics developments are making DNA sequencing routine and affordable – the backbone of much future research and application in healthcare and many other fields.

Summary

- Genes can be manipulated between a wide range of species – bacteria, fungi, animals, and plants.

- Restriction enzymes cut the donor DNA at specific sites. The relevant fragment is inserted into a vector that introduces the foreign DNA into cells of the host species.

- Recombinant DNA technology has a wide spectrum of uses. These include producing useful proteins in large quantities, producing transgenic animals and plants and human gene therapy.

- The polymerase chain reaction produces multiple copies of a DNA sequence.

- The standard method used to determine the nucleotide sequence of a DNA segment is Sanger's chain termination method (Sanger sequencing).

Problems

1. What roles do the following play in cloning a DNA fragment?

    (a) Restriction enzymes.

    (b) Vectors.

    (c) DNA ligase.

    (d) Electroporation.

2. In a PCR cycle, what processes typically occur at:

    (a) 72°C?

    (b) 95°C?

    (c) 55°C?

3. What are the advantages of quantitative (real-time) PCR over traditional PCR?

4. What is the role of ddNTPs in the chain termination method of DNA sequencing?

5. What type of gene cloning uses Ti plasmids?

6. What are the main stages in producing a transgenic mouse?

7. You hope to have successively cloned a gene of interest into the *lacZ* gene of a plasmid containing an tetracycline-resistance gene. You have introduced your plasmids into bacterial host cells sensitive to tetracycline. You plate out your bacteria on media containing both tetracycline and X-gal. Which of the following results would indicate recombinant clones?

(a) Cells that did not grow on the tetracycline/X-gal plates.

(b) Cells that form blue colonies on the tetracycline/X-gal plates.

(c) Cells that form white colonies on the tetracycline/X-gal plates.

8. Which of the following vectors and host cells would be the best choice to clone a eukaryotic gene?

(a) A bacterial plasmid/*E. coli*.

(b) A bacterial plasmid/yeast.

(c) A BAC/yeast.

(d) A YAC/*E. coli*.

(e) A YAC/yeast.

9. When cloning, is it better to use restriction enzymes that produce blunt ends or sticky ends?

10. Which of the following is not needed for a PCR reaction: *Taq* polymerase, ddNTPs, primers, template DNA?

11. Why are heat-stable DNA polymerases from thermophilic bacteria required for the PCR?

Solutions to Problems

Chapter 2

1. Let **B** = allele for black eyes and **b** = allele for red eyes.

   (a) F_1 all black-eyed. Black-eyed male (**BB**) × red-eyed female (**bb**) produces all black-eyed (**Bb**).

   (b) Three black-eyed mice to every one red-eyed mouse. F_1 black-eyed (**Bb**) × F_1 black-eyed (**Bb**) produces 1 **BB** (black-eyed) : 2 **Bb** (black-eyed) : 1 **bb** (red-eyed).

2. Perform a test cross (i.e. cross the black-eyed mouse with a red-eyed one). The presence of red-eyed mice among the test cross progeny would indicate the F_2 black-eyed mouse was heterozygous.

   If black-eyed mouse was **BB**, then **BB** × **bb** produces **Bb** (black eyed).

   If black-eyed mouse was **Bb**, then **Bb** × **bb** produces 50% **Bb** (black eyed) and 50% **bb** (red eyed).

3. (a) The results suggest that hair length is determined by a single gene with two alleles – one allele promoting the growth of short hair and the other long hair. The approximate 3 : 1 offspring ratio of short- to long-haired guinea pigs indicates that short hair is the dominant trait.

   (b) Heterozygous, **Ss** (if **S** = allele for short hair and **s** = allele for long hair). A 3 : 1 offspring ratio indicates heterozygous parents.

4. The results suggest that polydactyly is determined by a dominant allele at a single locus.

5. The F_1 poppies were spotted. A 3 : 1 offspring ratio indicates spotted is the dominant trait and that F_1 parents are heterozygous.

6. 100%; all their children will have dimples. Dimpled father (**DD**) × mother lacking dimples (**dd**) produces all children of genotype **Dd**.

7. The 3 : 1 ratio among the progeny produced by crossing flies Q and R indicates that one gene with two alleles determines body color with the gray body allele dominant to the black body allele. Flies Q and R are therefore heterozygous, while fly P is homozygous dominant. Thus, if **G** = allele for gray body and **g** = allele for black body, the results of crossing the three flies with a black-bodied fly are:

   Fly P (**GG**) × black-bodied fly (**gg**) produces 100% gray-bodied flies (**Gg**).

Fly Q or fly R (**Gg**) × black-bodied fly (**gg**) produces 50% gray-bodied flies (**Gg**) and 50% black-bodied flies (**gg**).

8. 480 plants should be large-leaved and 160 plants small-leaved. The F_1 results indicate that large leaves is the dominant trait. The F_1 plants are heterozygous. Crossing two heterozygous individuals produces a progeny ratio of 3 : 1 dominant to recessive individuals.

9. Nil, because neither blue-eyed parent has an allele for brown eyes.

10. Shell thickness is controlled by a single gene with two alleles: one allele (**T**) promotes the development of thick shells and the other allele (**t**) promotes the development of thin shells. Thus:

    Tree A (**Tt**) × thin-shelled tree (**tt**) produces 50% **Tt** (thick shelled) and 50% **tt** (thin shelled).

    Tree B (**TT**) × thin-shelled tree (**tt**) produces 100% **Tt** (thick shelled).

11. (a) Heterozygous, **Pp**. Polled (**Pp**) × polled (**Pp**) produces 1 **PP** (polled) : 2 **Pp** (polled) : 1 **pp** (horned).

    (b) (i) 3/4 or 75% (genotypes **PP** and **Pp**); (ii) 1/4 or 25% (genotype **pp**); (iii) 1/4 or 25% (genotype **PP**).

Chapter 3

1. Let **R** = allele for red flowers and **W** = allele for white flowers. Genotypes: red-flowering plants = **RR**, pink-flowering plants = **RW**, and white-flowering plants = **WW**.

    (a) **RW** (pink) × **RW** (pink) produces 1 **RR** (red) : 2 **RW** (pink) : 1 **WW** (white).

    (b) **WW** (white) × **RW** (pink) produces 1 **WW** (white) : 1 **RW** (pink).

    (c) **WW** (white) × **RR** (red) produces all **RW** (pink).

    (d) **WW** (white) × **WW** (white) produces all **WW** (white).

2. Henrik's parents have the genotypes $I^A i$ and $I^B i$. As Henrik is blood group B, his genotype must be $I^B I^B$ or $I^B i$; thus, at least one parent has an I^B allele. As he has a sister who is blood group A and a brother who is blood group 0, this indicates that one parent has an I^A allele and both must have an I^i allele.

3. Expected progeny phenotypes will be 1 yellow : 2 cream : 1 white coated (i.e. $C^Y C^Y \times C^W C^Y$ produces 1 $C^Y C^Y$: 2 $C^W C^Y$: 1 $C^W C^Y$).

4. Yes, 50% of F_2 plants will be expected to produce oval-shaped radishes. [F_1 plants $S^L S^R$. F_2 plants 25% $S^L S^L$ (long), 50% $S^L S^R$ (oval), and 25% $S^R S^R$ (round).]

5. No, because Charlie Chaplin could not have been the father. A child of blood group B has the genotype $I^B I^B$ or $I^B i$. No I^B allele could have been inherited from Joan Barry as she was blood group A ($I^A I^A$ or $I^A i$). Thus, the I^B allele came from the child's father. As Charlie Chaplin was blood group 0, his genotype was $i i$.

6. Seven offspring would be expected to have normal legs. Creeper cockerel (**Cc**) × creeper hen (**Cc**) produces 1 **CC** (lethal) : 2 **Cc** (creepers) : 1 **cc** (normal). Thus, surviving offspring are in the ratio of 2 creepers : 1 normal.

7. (a) Migratory behavior seems to be determined by one gene with two incompletely dominant alleles.

   (b) Let **M** = migratory allele and **N** = non-migratory allele. F_1 birds (**MN**) × non-migratory birds (**NN**) would be expected to produce 50% birds showing weak migratory behavior (**MN**) and 50% non-migratory birds (**NN**).

8. (a) All lentil seeds show the M1 marbled pattern.

   (b) 50% seeds show the M2 marbled pattern, 25% dotted, and 25% clear.

   (c) 50% seeds show M1 marbled pattern, 25% M2 marbled pattern, and 25% dotted.

9. Two platinum foxes to every one silver-coated fox.

   Platinum (**Pp**) × platinum (**Pp**) produces 1 **PP** (lethal) : 2 **Pp** (platinum) : 1 **pp** (silver).

10. It would appear that the three coat colors are controlled by one gene with two incompletely dominant alleles, C^R and **C**:

Parents	F_1 progeny	Parental genotypes
Cremello × palomino	50% cremello; 50% palomino	$C^R C^R \times C^R C$
Chestnut × palomino	50% chestnut; 50% palomino	$CC \times C^R C$
Palomino × palomino	25% cremello; 50% palomino; 25% chestnut	$C^R C \times C^R C$

11. (a) Recessive.

    (b) If **D** = allele for normal hearing and **d** = allele for deafness: (i) **DD** (has a child with a woman whose family have the gene, thus unlikely to be carrying the deaf allele); (ii) **dd**; (iii) **Dd**; (iv) **dd**.

12. (a) 1/4 or 25%.

    (b) 100%.

    (c) 27/64 (3/4 × 3/4 × 3/4).

Let **C** = cloven-footed allele and **c** = mule-footed allele. All F_1 pigs are cloven footed (**Cc**); 3/4 F_2 pigs are cloven footed (**CC** or **Cc**) and 1/4 are mule-footed.

13. Let **B** = allele for black feathers and **W** = allele for white feathers.

Erminette cockerel (**BW**) × erminette hen (**BW**) produces chicken in the expected proportions of 1 black (**BB**) : 2 erminette (**BW**): 1 white (**WW**).

(a) 1/16 (1/2 × 1/2 × 1/2 × 1/2, because each egg has a probability of 1/2 of being erminette).

(b) 1/256 (1/4 × 1/4 × 1/4 × 1/4, because each egg has a probability of 1/4 of being white).

(c) 1/256 (1/4 × 1/4 × 1/4 × 1/4, because each egg has a probability of 1/4 of being white, and also of being black).

14. 9/16 (3/4 × 3/4).

Chapter 4

1. Double heterozygote or **PpHh** (if **P, p, H**, and **h** = alleles for purple stemmed, green stemmed, hairy stemmed, and hairless stemmed, respectively). The approximate 9 : 3 : 3 : 1 proportions of the four offspring phenotypes indicate the heterozygous genotype.

2. 3/16 of the progeny. Do a Punnett square to show the possible fertilizations between gametes produced by two heterozygotes. Any individual of genotype **bbS-** will have brown hair and a short tail.

3. (a) The approximate 9 : 7 ratio of normal-sighted to blind crickets indicates complementary gene action between two loci. At least one dominant allele must be present at each locus to ensure normal sight.

(b) 1/16; individuals with the double homozygous dominant genotype, **AABB** (**A/a** and **B/b** represent the two loci and their alleles).

(c) 3/4 blind and 1/4 sighted crickets. Cross was **AaBb** × **aabb**, which produced 1 **AaBb** (sighted) : 1**aaBb** (blind): 1 **Aabb** (blind) : 1 **aabb** (blind).

4. If **P** = polled allele, **p** = horned allele, **R** = red hair allele, and **W** = white hair allele:

RWpp = genotype of the roan, horned bull

WWPP = genotype of the white, polled cow

RWpp × **WWPP** produces 50% **RWPp** (roan, polled) and 50% **WWPp** (white, polled).

5. Genotype of six bees that uncapped and removed infected larvae = **uurr**.

   Genotype of 20 bees that only uncapped = **uuR-**.

   Genotypes of 74 bees that exhibited no nest-cleaning behavior = **U-R-** and **U-rr**.

6. The approximate 9 : 7 F_2 ratio of plants with purple or colorless aleurone seed layers indicates complementary gene action between two loci. At least one dominant allele must be present at each in order for purple aleurones to be produced.

7. (a) 9/16 (plants of genotype **Ac-Li**).

   (b) 3/16 (plants of genotype **Ac-lili**).

   (c) 4/16 or 1/4 (plants of genotypes **acacLi-** and **acaclili**).

8. Parents: black coat, trotting gait (**BbTt**) and black coat, pacing gait (**Bbtt**):

Gametes	BT	Bt	bT	bt
Bt	BBTt black trotter	Bbtt black pacer	BbTt black trotter	Bbtt black pacer
bt	BbTt black trotter	Bbtt black pacer	bbTt chestnut trotter	bbtt chestnut pacer

9. Let **A** = allele for achondroplasia, **a** = allele for normal growth, **S** = allele for short sightedness, and **s** = allele for normal sight.

   Parents: **Aass** (achondroplasia man) × **aaSs** (short-sighted woman) produces 1 **AaSs** (achondroplasia, short-sighted) : 1 **aaSs** (normal growth, short-sighted) : 1 **Aass** (achondroplasia, normal-sighted) : 1 **aass** (normal growth and sight).

   (a) 1/2 (1/4 + 1/4).

   (b) 1/4.

10. The breeder's friend is suggesting the existence of dominant epistasis to explain the results of the three matings between the two white cats. This can be recognized by a 12 : 3 : 1 phenotype ratio among the progeny produced by a mating between two heterozygotes. Let **W** = suppressing allele, **w** = non-suppressing allele, **T** = tabby allele, and **t** = black allele.

    WwTt × **WwTt** produces 12 **W-T-** and **W-tt** : 3 **wwT-** : 1 **wwtt**.

Calculating a value for χ^2:

Phenotype	Observed numbers (O)	Expected numbers (E)	$O - E$	$(O - E)^2$	$(O - E)^2/E$
White	17	18	−1	1	0.06
Tabby	5	4.5	0.5	0.25	0.05
Black	2	1.5	0.5	0.25	0.17
Total	24	24			**0.28**

- Null hypothesis: there is no difference between observed results and an expected ratio of $12 : 3 : 1$

- Significance level = 0.05; degrees of freedom = 2; calculated χ^2 value = 0.28; critical χ^2 value = 5.99

As the calculated χ^2 value is less than the critical χ^2 value, the null hypothesis can be accepted. The friend's suggestion was correct.

Chapter 5

1. (a) Meiosis I; (b) mitosis, meiosis I, and meiosis II; (c) mitosis and meiosis II; (d) mitosis and meiosis II.

2.

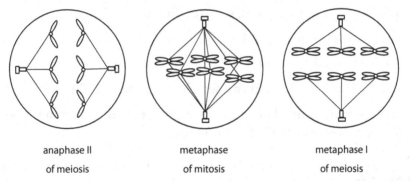

anaphase II	metaphase	metaphase I
of meiosis	of mitosis	of meiosis

3.

Mitosis	Meiosis
One division	Two divisions
Two daughter cells produced	Four daughter cells produced
Each daughter cell has the same number of chromosomes as the parental chromosome	Each daughter cell has half the number of chromosomes as the parental chromosome
No pairing of homologous chromosomes	Pairing of homologous chromosomes during prophase I
At anaphase sister chromatids separate	At anaphase I pairs of homologous chromosomes separate, while at anaphase II sister chromatids separate

Mitosis	Meiosis
No swopping of genetic information between chromosomes	Swopping of genetic information between non-sister chromatids of a chromosome pair during prophase I
Daughter nuclei genetically identical to parental nucleus	Daughter nuclei genetically different from parental nucleus
Supplies cells for growth, repair, and replacement	Supplies cells for sexual reproduction

4. (a) (i) A pair of chromosomes, one derived from an individual's father and the other from the mother, which are identical in their arrangement of genes. (ii) Two chromosomes formed by replication of a chromosome during S phase of the cell cycle and held together by their centromeres.

   (b) The arrangement of genes is the same on homologs and sister chromatids. Different alleles may be found at the same loci on each homolog, while identical alleles will be found at the same loci on sister chromatids.

5. Eight. The number of possible chromosome arrangements at the equator during metaphase I is $2^n - 1$, where n is the number of chromosome pairs. A diploid cell with eight chromosomes possesses four pairs of homologous chromosomes.

6. (a) Diploid, so 12; (b) haploid, so 6; (c) diploid, so 12.

7. (a) (iii); (b) (i); (c) (ii); (d) (iv).

8. (a) Cell 2; (b) cell 1; (c) cell 3.

9. (a) 24; (b) 46 autosomes and two sex chromosomes; (c) 24.

10. 54 chromosomes (the Shetland pony egg contained 32 chromosomes and the zebra sperm contained 22 chromosomes).

Chapter 6

1. (a) 1/2; (b) 1/4; (c) 1/2.
   $X^A X^a \times X^A Y$ produces $X^A X^A$, $X^A X^a$, $X^A Y$, and $X^a Y$ ($X^a Y$ shows trait).

2. Both parents are likely to be normal sighted with the genotypes: mother $X^C X^c$ and father $X^C Y$.
   $X^C X^c \times X^C Y$ produces $X^C X^C$, $X^C X^c$, $X^C Y$, and $X^c Y$. Both sons were $X^c Y$ and color blind.

3. 1/4. Woman $X^H X^h$ (received hemophiliac allele from father) and man $X^H Y$ (he is normal).
   $X^H X^h \times X^H Y$ produces $X^H X^H$, $X^H X^h$, $X^H Y$, and $X^h Y$ (hemophiliac).

4. 50% black, 25% ginger, and 25% tortoiseshell. If B = black hair allele and G = ginger hair allele:

$X^B X^G$ (tortoiseshell female) × $X^B Y$ (black male) produces $X^B X^B$ (black female), $X^B Y$ (black male), $X^B X^G$ (tortoiseshell female), and $X^G Y$ (ginger male).

5. As her uncle suffered from muscular dystrophy, the woman could be a carrier for the disease-causing allele and therefore could pass the allele on to any children she might have. Only a son could express the condition. The probability of this occurring is 1/8.

   The uncle's mother (i.e. the woman's grandmother) was a carrier. The woman's mother, therefore, had a probability of 1/2 of being a carrier. *If* she was a carrier the woman had a probability of 1/2 of being a carrier. Thus, her probability of being a carrier is $1/2 \times 1/2 = 1/4$. If she *is* a carrier, then there is a probability of 1/2 that she has a son who has muscular dystrophy. Thus, there is an overall probability of $1/2 \times 1/2 \times 1/2 = 1/8$.

6. (a) 1/16 ($1/2 \times 1/2 \times 1/2 \times 1/2$ because each puppy has a probability of 1/2 of being male and also a probability of 1/2 of being female).

   (b) 1/16 ($1/2 \times 1/2 \times 1/2 \times 1/2$ because each puppy has a probability of 1/2 of being female).

   (c) 5/8 [$1/2 + (1/2 \times 1/2 \times 1/2)$].

7. Five. Barr bodies are condensed inactivated X chromosomes. All but one X chromosome is inactivated in each somatic cell.

8. Affected stallion. Let D = disease allele and d = normal allele.

   If the mare is affected: $X^D X^d \times X^d Y$ produces $X^D X^d$ (affected), $X^d X^d$, $X^D Y$ (affected), and $X^d Y$.

   If the stallion is affected: $X^d X^d \times X^D Y$ produces $X^D X^d$ (affected) and $X^d Y$.

9. Feather color is a sex-linked trait, with gray the dominant color. If G = gray allele and g = white allele; ZZ = male bird and ZW = female bird:

   $Z^G Z^g \times Z^G W$ produces $Z^G Z^G$, $Z^G Z^g$, $Z^G W$, and $Z^g W$ (white females).

   $Z^g Z^g \times Z^g W$ produces $Z^g Z^g$ and $Z^g W$ (white females).

10. If male parent is solid color and female parent barred, then all male chickens will be barred and all female chickens will be solid colored.

   $Z^b Z^b \times Z^B W$ produces $Z^B Z^b$ (barred male chickens) and $Z^b W$ (solid female chickens).

Chapter 7

1. Linkage of the two loci is indicated by the larger number of parental phenotypes (brown and albino) compared with the recombinants (orange and black). If the two genes were on separate chromosomes, the cross would have produced approximately equal numbers of the four phenotypes (i.e. a 1 : 1 : 1 : 1 ratio).

2. Let **R** = resistance allele, **r** = sensitivity allele, **G** = green pod allele, and **g** = orange pod allele.

   (a) F_1 phenotype is green pods and resistance to pea mosaic virus:

   F_1 genotype is $\quad \dfrac{\mathbf{R} \quad \mathbf{G}}{\mathbf{r} \quad \mathbf{g}}$

   (b) Possible phenotypes are green pods and resistance to virus; green pods and sensitivity to virus; orange pods and resistance to virus; and orange pods and sensitivity to virus. The first and fourth phenotypes will be the most frequent.

3. (a) The unequal proportions among the four progeny classes, with greater numbers of the parental phenotypes (colored fur, normal gait and waltzer, albinos).

   (b) If **C** = allele for colored fur, **c** = albino allele, **N** = normal gait, and **n** = waltzing gait, the genotypes producing each of the four phenotypes are:

   $$\dfrac{\mathbf{C} \quad \mathbf{N}}{\mathbf{c} \quad \mathbf{n}} \text{ colored, normal} \qquad \dfrac{\mathbf{C} \quad \mathbf{n}}{\mathbf{c} \quad \mathbf{n}} \text{ colored, waltzer}$$

   $$\dfrac{\mathbf{c} \quad \mathbf{N}}{\mathbf{c} \quad \mathbf{n}} \text{ albino, normal} \qquad \dfrac{\mathbf{c} \quad \mathbf{n}}{\mathbf{c} \quad \mathbf{n}} \text{ albino, waltzer}$$

   (c) 50% albino, waltzer and 50% albino, normal gait mice:

   $$\dfrac{\mathbf{c} \quad \mathbf{n}}{\mathbf{c} \quad \mathbf{n}} \times \dfrac{\mathbf{c} \quad \mathbf{N}}{\mathbf{c} \quad \mathbf{n}} \text{ produces } \dfrac{\mathbf{c} \quad \mathbf{n}}{\mathbf{c} \quad \mathbf{n}} \text{ and } \dfrac{\mathbf{c} \quad \mathbf{N}}{\mathbf{c} \quad \mathbf{n}}$$

4. Equal numbers of the two parental phenotypes; that is, the double recessive (produced by **aabb**) and the double dominant (produced by **AaBb**):

   $$\dfrac{\mathbf{a} \quad \mathbf{b}}{\mathbf{a} \quad \mathbf{b}} \times \dfrac{\mathbf{A} \quad \mathbf{B}}{\mathbf{a} \quad \mathbf{b}} \text{ produces } \dfrac{\mathbf{a} \quad \mathbf{b}}{\mathbf{a} \quad \mathbf{b}} \text{ and } \dfrac{\mathbf{A} \quad \mathbf{B}}{\mathbf{a} \quad \mathbf{b}}$$

5. Independent assortment of two genes can occur if the two are far apart on the same chromosome. Thus, during prophase I of meiosis, a chiasma always forms between the two genes, producing recombinant gametes.

6. 7.7 cM, because 1% recombination indicates a map distance of 1 cM.

7. 26 cM:

   $$\% \text{ recombinants} = \frac{\text{number of recombinants}}{\text{total number of progeny}} = \frac{21}{80} = 0.26$$

   Recombinant frogs are green, large padded and blue, normal padded.

8. The genes controlling color of body and shape of antennae are on the same chromosome. As only a few crosses yielded any recombinant progeny, flies with aristapedia antennae and gray bodies and normal antennae and ebony bodies, the two genes must be very close together.

9. (a) 860 **AaBb**, 140 **Aabb**, 140 **aaBb**, and 860 **aabb** (because the two genes are 14 cM apart, 14% of the offspring are expected to be recombinants, i.e. to possess the genotypes **Aabb** and **aaBb**).

   (b) The same results as (a).

10.

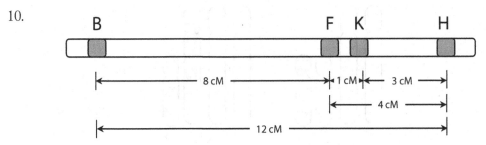

Chapter 8

1. (a) 45; (b) 69; (c) 47; (d) 48.

2. (a) Pericentric inversion; (b) duplication; (c) duplication and paracentric inversion; (d) translocation.

3. (a) Hybrid chromosome number = 29; allopolyploid chromosome number = 58.

   (b) Hybrid will be infertile: there are 29 different chromosomes and so homologous chromosome pairing is impossible during prophase I of meiosis. Unbalanced gametes result. In the allopolyploid, produced by chromosome doubling in the hybrid, each chromosome will now be represented twice, pairing is once again possible during meiosis and balanced gametes will be produced, containing two copies of each chromosome. Thus, fertile offspring can be produced.

4. Problems generally only result if the break point of an inversion or translocation occurs within a gene. However, meiosis within the cells of an individual containing an inversion or a translocation can result in unbalanced gametes, and so unbalanced zygotes at fertilization. Thus, problems can occur in subsequent generations.

5. 1/8 gametes would be expected to contain all three chromosomes, 3/8 gametes contain two chromosomes, 3/8 gametes contain one chromosome, and 1/8 gametes would be expected to contain no chromosome (considering the chromosome that is trisomic). This range of gametes results because of

pairing problems at prophase I of meiosis. Three chromosomes cannot successfully pair.

6. Gene order is **qsprut** (note that the deletions would be on just one of a pair of homologous chromosomes).

7. One of the parents' cells (parent 2 below) contains a balanced reciprocal translocation between chromosomes 9 and 12. An unbalanced gamete from this parent was used at fertilization, resulting in the chromosome composition of the child shown in the diagram.

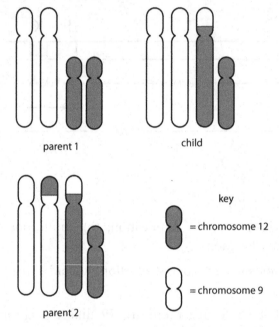

8. Twelve chromosomes per gamete. An autohexaploid has six chromosome sets in each cell, thus 12 chromosomes will be found in a haploid set.

9. An allotetraploid possesses four copies of each chromosome. Thus, if there are 44 chromosomes in each cell, there are 11 different chromosomes, or linkage groups.

10. Mother. A woman with Turner syndrome is XO. If she is color blind, her genotype is X^cO (c = color-blind allele). Thus, she must have inherited her father's X chromosome.

11. 64.

Chapter 9

1. (a) Mean; (b) distribution with the bigger variance; (c) 68.26% population lie within one standard deviation and 95.44% population lie within two standard deviations.

2.

Nest	Woodland weights (g)	$x - \bar{x}$	$(x - \bar{x})^2$	Scrubland weights (g)	$x - \bar{x}$	$(x - \bar{x})^2$
1	10.4	−0.11	0.012	9.01	−0.56	0.314
2	10.2	−0.31	0.096	9.86	0.29	0.084
3	10.6	0.11	0.012	9.71	0.14	0.02
4	9.4	−1.11	1.232	8.84	−0.73	0.533
5	11.0	0.49	0.24	9.33	−0.24	0.058
6	10.8	0.31	0.096	10.16	0.59	0.348
7	11.3	0.79	0.624	10.8	1.23	1.513
8	10.3	−0.21	0.044	8.94	−0.63	0.397
9	10.3	−0.21	0.044	9.66	0.09	0.008
10	10.8	0.31	0.096	9.43	−0.14	0.02

(a) Mean weight of (i) woodland nestlings is 10.51 g; (ii) scrubland nestlings is 9.57 g.

(b) Variance of (i) woodland nestlings = 0.277; (ii) scrubland nestlings = 2.947.

(c) Scrubland population has the larger standard deviation (note standard deviation = √variance): (i) standard deviation of woodland population = 0.527; (ii) standard deviation of scrubland population = 0.572.

3. F_2.

4. (a) $H^2 = 0.653$.

$H^2 = V_g/V_p$; $V_p = V_e + V_g$. F_1 variance represents V_e; F_2 variance represents V_p. Using F_1 and F_2 data: $4.96 = 1.72 + V_g$. $V_g = 4.96 - 1.72 = 3.24$. $H^2 = 3.24/4.96 = 0.653$.

(b) Yes.

5. (a) If the two varieties are subject to the same growth conditions, then individuals of the variety showing the low variance are likely to be genetically similar, while individuals of the other variety have a range of different genotypes. Alternatively, both varieties could be genetically homogeneous, but the variety showing the higher variance could be subjected to a wider range of growth conditions.

(b) The variety showing the lower variance, as you could be more certain of the range of fruit weights you would obtain.

(c) Variety with the higher variance, as you are more likely to get some heavier fruit to include in your breeding program.

6. (a) 2.28%: 16.4 min represents two standard deviations above the mean; 4.56% of children will complete the maze in a time three standard deviations or more above or below the mean (see Figure 9.2). Thus, 2.28% of children will take more than two standard deviations.

(b) 84.13%: 50% of children will reach the center in a time equal to or faster than the mean; 34.13% of children will reach the center in a time within one standard deviation below the mean; 6.8 min represents one standard deviation below the mean.

(c) 15.87%: all children not falling within the range of answer (a).

7. Heritability would decrease. After many generations of selective breeding, the population would become genetically uniform (i.e. possess genes for large berries).

8. (a) All F_1 pumpkins of weight 4.5 kg.

(b) F_2 pumpkins of 3, 3.5, 4, 4.5, 5, 5.5, and 6 kg in the ratio of $1 : 6 : 15 : 20 : 15 : 6 : 1$.

9. (a) $H^2 = 0.69$. $V_g = 4.2 + 1.6 + 0.3 = 6.1$. $H^2 = V_g/V_p = 6.1/8.8$.

(b) $h^2 = 0.48$. $h^2 =$ additive genetic variance $(V_a)/V_p = 4.2/8.8$.

10. (a) $h^2 = 0.5$, which also is determined by selection response (R)/selection differential (S): $R = 15.5$ cm $- 13$ cm $= 2.5$ cm; $S = 18$ cm $- 13$ cm $= 5$ cm. Thus, $h^2 = 2.5/5 = 0.5$ (see Box 9.3). A selective breeding program is likely to be effective in increasing the length of the 8-week salmon.

11. (a) $h^2 = 0.32$. $R/S = 8/25$. A moderate value for h^2 which indicates that a selective breeding program will have some impact on increasing average body weight.

Chapter 10

1. (a) 0.8 (note frequency of dominant allele $= p$, frequency of recessive allele $= q$; $p + q = 1$; as $q = 0.2$ and $p + q = 1$, then, $p = 0.8$).

(b) 0.32. Use the Hardy–Weinberg law which states that the three genotypes that can be produced by a monohybrid locus with two alleles present in a population are in the frequencies: p^2 (homozygous dominant), $2pq$ (heterozygote), and q^2 (homozygous recessive). So, using $2pq$, the heterozygote frequency is 2 $\times$ 0.8 $\times$ 0.2.

2. 188 vipers. Non-poisonous vipers possess a homozygous recessive genotype, vv. The frequency of allele **v**, or q, is 0.25. Thus, the frequency of the recessive genotype, or q^2, is $(0.25)^2 = 0.0625$. The population consists of 3000 vipers. Therefore, the number of non-poisonous vipers $= 3000 \times 0.0625 = 187.5$ (or 188 to the nearest whole number).

3. Assume that the population is in genetic equilibrium, thus allele and genotype frequencies will be the same in the next generation.

(a) 81%. In order to calculate frequencies of individuals showing the heterozygous and homozygous dominant genotypes, values for the dominant and recessive alleles must first be calculated; 81% of individuals are homozygous recessive or q^2. If $q^2 = 0.81$, $q = \sqrt{0.81} = 0.9$, thus $p = 0.1$.

(b) 18%. Heterozygote frequency $= 2pq = 2 \times 0.9 \times 0.1 = 0.18$.

(c) 1%. Frequency of homozygous dominant $= p^2 = (0.1)^2 = 0.01$.

4. (a) Phenotype frequency; (b) genotype frequency; (c) allele frequency.

5. 1 in 139. $q = 0.00362$; $p = 0.9964$. $2pq = 0.00721$, or 1 in 139 individuals.

6. (a) 0.43. $q^2 = 0.1$. Thus, $q = \sqrt{0.1} = 0.316$ and $p = 0.684$. $2pq$ (heterozygote frequency) $= 2 \times 0.316 \times 0.684 = 0.43$.

(b) 0.48

$$= \frac{\text{frequency of heterozygote}}{\text{frequency of individuals with dominant phenotype}}$$

$$= \frac{2pq}{p^2 + 2pq} = \frac{0.432}{0.468 + 0.432} = 0.48$$

7. 0.53. $0.78 = p^2 + 2pq$; thus $0.22 = q^2$; $q = \sqrt{0.22} = 0.47$. p (resistance allele frequency) $= 0.53$.

8. (a) Genetic drift, migration, mutation, selection.

(b) Recessive alleles are also present in the heterozygote. Selection will either not act against these heterozygotes, or at a reduced level.

(c) Much lower – most individuals are homozygous dominant or recessive.

(d) It is the only allele present at a given locus, which is therefore monomorphic.

(e) As mutations generally occur at an extremely low frequency, of the order of 1×10^{-5} per generation. Thus, it takes many generations for a substantial change in allele frequency to occur.

(f) Small population size.

9. (a) and (b):

Phenotype	Number of individuals	Genotype	Genotype frequency	Number of alleles	
				B	**G**
Green	120	**GG**	0.6 (120/200)	240	
Brownish-green	60	**BG**	0.3 (60/200)	60	60
Brown	20	**BB**	0.1 (20/200)		40
Total	200			300	100

$$\text{Frequency allele } \mathbf{B} = \frac{300}{400} = 0.75$$

$$\text{Frequency allele } \mathbf{G} = \frac{100}{400} = 0.25$$

(c) Expected frequencies $= p^2, 2pq,$ and q^2. Using the above values, the expected frequencies are: **GG** = 0.562, **BG** = 0.375, and **BB** = 0.0625.

(d) Perform a χ^2 test:

Phenotype	Observed numbers (O)	Expected number (E)	$O - E$	$(O - E)^2$	$(O - E)^2/E$
Green	120	112.5	7.5	56.25	0.5
Brownish-green	60	75	−15	225	3.0
Brown	20	12.5	−7.5	56.25	4.5
Total	200	200			**8.0**

- Null hypothesis: there is no difference between observed and expected numbers

- Significance level = 0.05; degrees of freedom = 2; calculated χ^2 value = 8.00; critical χ^2 value = 5.99

As the calculated χ^2 value is greater than the critical χ^2 value, at a significance level of 0.05 and 3 degrees of freedom, the null hypothesis is rejected. The population of frogs is *not* in Hardy–Weinberg equilibrium when the color locus is examined.

10. Use the same method as outlined in the answer to Question 9.

Frequency of allele **A** = 0.56 and frequency of allele **B** = 0.44.

Expected genotype frequencies: **AA** = 0.31, **AB** = 0.49, and **BB** = 0.19.

To test for Hardy–Weinberg equilibrium, perform a χ^2 test.

- Null hypothesis: there is no difference between observed and expected numbers

- Significance level = 0.05; degrees of freedom = 2; calculated χ^2 value = 12.4; critical χ^2 value = 5.99

As the calculated χ^2 value is greater than the critical χ^2 value, at a significance level of 0.05 and three degrees of freedom, the null hypothesis is rejected. The population of salamanders is *not* in Hardy–Weinberg equilibrium when the malate dehydrogenase locus is examined.

Chapter 11

1. (a) 400; (b) 200; (c) 45; (d) 400; (e) 155. If there are 200 nucleotide pairs in a fragment of DNA, there must be a total of 400 bases. If 45 of these are thymine, 45 will be adenine, because of complementary base pairing. Together, these account for 90 of the 400 bases. The remaining 310 bases are cytosine and guanine, 155 of each.

2. (a) 3′-TGGCCATCTTAGC-5′; (b) 5′ → 3′.

3. 23% are adenine. If there are 27% guanine bases, there are also 27% cytosine bases. Together these account for 54% of the bases; 46% are adenine and thymine, 23% of each.

4. Opposite directions (note that the two strands of DNA are anti-parallel).

5. (a) 30,000 complete turns; (b) 300,000 nucleotide pairs. One complete turn of the DNA double helix occupies 3.4 nm and one nucleotide pair occupies 0.34 nm.

6. (b) and (c). Substitute some figures (e.g. 20% A, 20% T, 30% C, and 30% G) and work out the ratios.

7. (a) False; (b) false (complementary, not identical); (c) true (because of complementary base pairing); (d) false; (e) false (there will be 16% cytosine).

8. These terms refer to modes of DNA replication. Semi-conservative replication refers to the fact that each new DNA molecule possesses one conserved parental strand and one newly synthesized strand. Conservative replication refers to the suggestion in the 1950s that replication might produce one molecule with both parental strands and the other with both strands newly synthesized.

9. DNA polymerases can only synthesize in a 5′ → 3′ direction. Owing to the anti-parallel nature of the two DNA strands, if replication proceeds in one overall direction, only one new strand can be synthesized in a 5′ → 3′ direction.

The other strand is synthesized in segments in the apparently wrong direction and the segments subsequently joined together.

10. 39% of the bases are also thymine. Thus, 11% are cytosine and 11% are guanine. The DNA, therefore, contains an unusually high percentage of adenine and thymine.

11. (a) 344,560 nucleotides; (b) 58.58 μm (a base pair occupies 0.34 nm, so 172,280 × 0.34 nm).

Chapter 12

1.

DNA	RNA
Pentose sugar: deoxyribose	Pentose sugar: ribose
Pyrimidines: cytosine and thymine	Pyrimidines: cytosine and uracil
Double stranded	Single stranded
One molecule contains millions of nucleotides	One molecule contains hundreds or thousands of nucleotides
One function: contains encoded information	A variety of functions (e.g. mRNA carries encoded information of one gene; tRNA carries amino acids; rRNA is a constituent of ribosomes)
Located in the nucleus and two organelles, the mitochondria and chloroplasts	Made in the nucleus, but then found in the cytoplasm: all three main types also found in mitochondria and chloroplasts

2. (a) tRNA; (b) mRNA, tRNA, and rRNA; (c) tRNA; (d) mRNA; (e) mRNA; (f) tRNA.

3. (a) 3′-**CCUUGGGUC**-5′; (b) CAG.

4. Deletion of a base changes the reading frame. All codons after the mutation point are altered; thus, all encoded amino acids are different and so the resulting protein is rendered non-functional. By contrast, a substitution, which swops one base for another, does not alter the reading frame.

5. Transcription: alignment of ribonucleotides against the template DNA strand. Translation: tRNA anticodon binding to mRNA codon.

6.

mRNA	tRNA
Single strand of RNA	Formed as a single strand, then folds into a cloverleaf structure with double-stranded regions
Main bases: ATGC	Contains additional bases to ATGC
Contains the encoded genetic information	Binds to and transports amino acids from the cytoplasm to ribosomes during translation

7. Nucleotides, codons, exons, genes, chromosomes, genomes.

8. 582 nucleotides: $193 \times 3 = 579$ nucleotides (three nucleotides encode one amino acid) + 3 nucleotides of a stop codon.

9. 1 C; 2 J; 3 G; 4 I; 5 H; 6 A; 7 E; 8 D; 9 F; 10 B.

10.

T	C	G	A	C	C	T	G	A	C	T	T	DNA
A	G	C	T	G	G	A	C	T	G	A	A	
A	G	C	U	G	G	A	C	U	G	A	A	mRNA transcribed
U	C	G	A	C	C	U	G	A	C	U	U	tRNA anticodon
Serine			Tryptophan			Threonine			Glutamine			Encoded amino acid

Chapter 13

1. (a) The same restriction enzyme is used to make a cut in the vector and in the DNA to be cloned.

(b) The vector (e.g. plasmid) carries the DNA to be cloned into the host cell (e.g. bacteria).

(c) DNA ligase catalyzes the sealing of the joint between the cloned DNA fragment and vector DNA.

(d) Electroporation enables transformation: host cells are treated with a brief, weak electric shock that temporarily permeabilizes the cell membranes. This allows vectors to enter cells.

2. (a) Extension (i.e. *Taq* DNA polymerase catalyzes addition of dNTPs to the primer).

(b) Denaturation of the two DNA strands.

(c) Annealing of primers to their target sequences.

3. Quantitative real-time PCR is more sensitive than traditional PCR. It is quicker – no post-PCR analysis of DNA is needed as DNA is detected as it is synthesized. DNA produced by real-time PCR can be quantified. As all reactions occur in a single tube during real-time PCR there is less chance of contamination.

4. ddNTPs are a component of the PCR reaction generating DNA fragments for sequencing. When *Taq* DNA polymerase adds a ddNTP, this terminates DNA synthesis. Addition of a ddNTP is random. The results are PCR fragments of

all possible different lengths, each terminated by a fluorescing ddNTP that is detected by the sequencing machine.

5. Ti plasmids carry genes into plant genomes, using the bacterium *Agrobacterium tumefaciens* as the host cell.

6. Recombinant DNA technology is first used to build the DNA vector to insert into the mouse. Along with the gene of interest and vector DNA are the promoter sequences that enable expression of the gene in the mouse. The vector is injected into the nucleus of newly fertilized eggs that are grown for a few generations. Viable embryos are then implanted in pregnant females.

7. (c).

8. (e).

9. It is better to use restriction enzymes that produce sticky ends. This produces single-stranded ends to both the cut cloning vector and DNA insert. These single strands hybridize to each other by complementary base pairing and so fix the DNA insert in the vector.

10. ddNTPs are not needed. These are specialized dNTPs that are used in the PCR reaction that generates DNA fragments for sequencing.

11. The temperature of the PCR reaction mix is raised to 95°C for DNA denaturation. Normal enzymes are also denatured by this high temperature. Thermophilic bacteria that permanently live in environments with high temperatures have evolved enzymes that are heat stable.

Appendix

$$(\chi^2 \text{ critical values})$$

Degrees of freedom (d.f.)	Probability level of significance			
	0.1	0.05	0.025	0.01
1	2.71	3.84	5.02	6.64
2	4.61	5.99	7.38	9.21
3	6.25	7.82	9.35	11.35
4	7.78	9.49	11.14	13.28
5	9.24	11.07	12.83	15.09
6	10.65	12.59	14.45	16.81
7	12.02	14.07	16.01	18.48
8	13.36	15.51	17.54	20.09
9	14.68	16.92	19.02	21.67
10	15.99	18.31	20.48	23.21
11	17.28	19.68	21.92	24.73
12	18.55	21.03	23.34	26.22
13	19.81	22.36	24.74	27.69
14	21.06	23.69	26.12	29.14
15	22.31	25.00	27.49	30.58
16	23.54	26.30	28.85	32.00
17	24.77	27.59	30.19	33.41
18	25.99	28.87	31.53	34.81
19	27.20	30.14	32.85	36.19
20	28.41	31.41	34.17	37.57
21	29.62	32.67	35.48	38.93
22	30.81	33.92	36.78	40.29
23	32.01	35.17	38.08	41.64
24	33.20	36.42	39.36	42.98
25	34.38	37.65	40.65	44.31
26	35.56	38.89	41.92	45.64
27	36.74	40.11	43.20	46.96
28	37.92	41.34	44.46	48.28
29	39.09	42.56	45.72	49.59
30	40.26	43.77	46.98	50.89
31	41.42	44.99	48.23	52.19
32	42.59	46.19	49.48	53.49
33	43.75	47.40	50.73	54.78
34	44.90	48.60	51.97	56.06
35	46.06	49.80	53.20	57.34

Glossary

Addition rule: the probability of one event or another event occurring is calculated by adding the probabilities of each separate event.

Adenine: a purine base which pairs with thymine in DNA and uracil in RNA.

Allele: one of two or more forms of a gene.

Alloploid: an individual, or cell, which possesses two or more distinct chromosome sets derived from different species.

Amino acids: the covalently linked building blocks of proteins.

Amphidiploid: an organism with a diploid set of chromosomes derived from each parent.

Anaphase: the stage of nuclear division during which homologous chromosomes or sister chromatids separate and move towards opposite poles of the cell.

Aneuploid: an individual, or cell, in which the chromosome number is not an exact multiple of the haploid set.

Anticodon: the nucleotide triplet in a tRNA molecule that is complementary to an mRNA codon.

Antisense strand: the template strand during transcription. It has the complementary sequence to mRNA.

Aster: a star shaped cluster of microtubules which radiate from the centrioles during mitosis and meiosis.

Autopolyploid: an individual, or cell, which possesses three or more identical sets of chromosomes.

Autosome: any chromosome other than a sex-determining chromosome.

Backcross: a cross that involves an F_1 heterozygote and one of the parents (generally the recessive parent).

Bacterial artificial chromosome (BAC): a cloning vector, derived from the F plasmid, into which 100–300 kb of DNA can be inserted.

Bacteriophage: a virus that infects bacteria.

Barr body: an inactive X chromosome, visible as a densely staining mass within the somatic nucleus of mammalian females.

Binary fission: a form of asexual reproduction during which a parent cell divides into two identical daughter cells.

Bioinformatics: the use of computers to store, model and analyse biological data.

Bivalent: a pair of associated homologous chromosomes during prophase I of meiosis.

Blunt end: the end of a DNA fragment in which there are no unpaired bases.

Bottleneck: a transitory shrinking of population size when alleles may be lost from the gene pool.

Carrier: an individual who is heterozygous for a recessive trait.

Cell cycle: the sequential phases of growth of an individual cell; starting with G_1 (gap 1) followed by S (DNA synthesis), G_2 (gap 2), and, finally, M (mitosis).

Central dogma: the idea that the encoded information in DNA flows via RNA into protein.

Centimorgan (cM): a unit of distance between genes on a chromosome; 1 cM corresponds to a recombination frequency of 1%.

Centriole: a structure, consisting of a collection of tiny microtubules, around which the spindle is organized during mitosis and meiosis.

Centromere: specialized region of a chromosome to which the spindle fibers attach during nuclear division.

Checkpoint: A cell cycle process that controls cell cycle progression by monitoring the successful completion of specific cell cycle events.

Chiasma: the location of a crossover between non-sister chromatids during prophase I of meiosis.

Chromatid: one of the two longitudinal subunits of a replicated chromosome, joined to its sister chromatid at the centromere.

Chromatin: the complex of histone proteins and DNA that make up chromosomes.

Chromosome: a thread-like structure consisting of chromatin, containing genetic information arranged in a linear sequence.

Co-dominance: condition in which the phenotypic effects of a pair of alleles are fully and simultaneously expressed in a heterozygote.

Codon: a triplet of nucleotides in a DNA or RNA molecule that specifies an amino acid.

Colchicine: an alklaloid extracted from autumn crocus which inhibits spindle formation and so arrests cells in metaphase of mitosis.

Collinearity: the parallel relationship between the linear sequence of codons in deoxyribonucleic acid and the order of amino acids in the polypeptide product that they specify.

Complementary base pairing: a chemical affinity between nitrogenous bases, such that adenine pairs only with thymine and guanine pairs only with cytosine.

Concordance: the presence of the same trait in a pair of twins.

Cosmid: a hybrid cloning plasmid consisting of DNA from a plasmid and from lambda phage.

Coupling: a term used during gene mapping to describe all dominant alleles linked together on one chromosome and all recessive alleles on the partner homologous chromosome.

Crossing over: the exchange of genetic material between homologous chromosomes during prophase I of meiosis; it produces recombination.

Cytokinesis: cell division that follows mitosis and meiosis.

Cytosine: a pyrimidine base which pairs with guanine in DNA and RNA.

Deletion: the loss of chromosomal material, ranging from a single nucleotide to many genes.

Deoxyribonucleotide (dNTP): a purine or pyrimidine base bonded to deoxyribose which is linked to a phosphate.

Dideoxyribonucleotide (ddNTP): a deoxyribonucleotide whose deoxyribose lacks a 3′-OH group. It terminates a growing nucleotide chain.

Dihybrid cross: a genetic cross in which the parents possess different forms of two traits.

Dioecious: male and female flowers are on separate plants.

Diploid: an individual, or cell, which possesses two copies of each chromosome.

Discordance: a pair of twins do not share the same trait.

DNA (deoxyribonucleic acid): a macromolecule consisting of two anti-parallel polynucleotide chains held together by hydrogen bonds; the primary carrier of genetic information.

DNA library: a collection of cloned DNA fragments from one source.

DNA ligase: an enzyme that catalyzes the formation of a phosphodiester bond.

DNA polymerase: an enzyme that catalyzes the synthesis of DNA from deoxynucleotides and a DNA template.

Dominance: a condition in which the phenotypic effect of only one of a pair of alleles is expressed in a heterozygote.

Double crossover: two separate exchanges of genetic material between homologous chromosomes during prophase I of meiosis.

Double helix: describes the helical configuration of the two anti-parallel polynucleotide chains of DNA.

Duplication: the presence of a repeated segment of a chromosome.

Equator: the midpoint between two centrioles in a cell undergoing mitosis or meiosis. It is where pairs of chromatids or of homologous chromosomes align during metaphase.

Enzyme: a protein that catalyzes a specific biochemical reaction.

Epistasis: a non-reciprocal interaction between genes such that an allele at one gene interferes with the expression of alleles at another gene.

Euchromatin: lightly staining chromosomal regions that are transcriptionally active during interphase.

Eukaryote: an organism whose cells possess a true nucleus and membranous organelles.

Euploid: an individual, or cell, with one or more complete sets of chromosomes.

Exons: the coding portions of genes that are transcribed and translated into protein.

F_1 (first filial) generation: the progeny resulting from the first cross in a series.

F_2 (second filial) generation: the progeny resulting from a cross involving the F_1 generation.

Fixation: a condition in which all members of a population are homozygous for a given allele.

Fluorescence *in situ* hybridization (FISH): a technique that uses fluorescent probes to detect the presence (or absence) of specific DNA sequences on a chromosome.

Founder effect: the establishment of a population by a small group of individuals whose genotypes carry only a fraction of the different alleles in the parental population; genetic drift acts to alter allele frequencies.

Frameshift mutation: the insertion or removal of one or more nucleotides into or from a gene, which shifts the codon reading frame in all codons following the change.

Gamete: the haploid reproductive cell (sperm or ovum).

Gametophyte: the haploid gamete producing generation of plants.

Gene: the fundamental unit of heredity that occupies a specific chromosomal locus.

Gene flow: the gradual exchange of genes between two populations brought about by the dispersal of gametes or the migration of individuals.

Gene pool: a common set of genes, and their alleles, shared by a group of interbreeding individuals.

Gene therapy: the treatment of a disease caused by a malfunctioning gene by inserting a normal gene into the cells of the affected organism.

Genetic drift: the random variation in allele frequency from generation to generation; most often observed in small populations.

Genetic equilibrium: the maintenance of allele frequencies at a constant value in successive generations, if no outside forces are acting on them.

Genome: the totality of an organism's DNA.

Genomics: the discipline of genetics concerned with the structure, function, evolution, and mapping of genomes.

Genotype: the full complement of an individual's genes *or* the allelic composition of one or a few genes.

Germ line: gamete-producing cells.

Giemsa stain: a type of stain that produces G bands in chromosomes.

Guanine: a purine base which pairs with cytosine in DNA and in RNA.

Haploid: an individual, or cell, that possesses one copy of each chromosome.

Hardy–Weinberg law: the principle that allele and genotype frequencies will remain constant in succeeding generations in an infinitely large population and in the absence of mutation, migration, selection, and non-random mating.

Hemizygous: a condition in which only a single copy of a gene is present; usually applies to genes on the X chromosomes in heterogametic males.

Heritability: the proportion of the variability of a trait attributable to genetic factors.

Heterochromatin: heavily staining chromosomal regions that are condensed and transcriptionally inactive.

Heterogametic sex: sex that produces gametes containing different sex chromosomes.

Heterozygote: an individual with two different alleles at a locus.

Histone: the protein around which DNA is wound in a chromosome.

Homogametic sex: sex that produces gametes containing identical sex chromosomes.

Homologs: chromosomes that possess identical loci and pair during meiosis.

Homozygote: an individual with two identical alleles at a locus.

Human Genome Project: an international scientific project to determine the complete DNA sequence of the human genome, to identify all genes and understand their function.

Hybridization: the formation of a double-stranded nucleic acid molecule through mixing two single strands of DNA and/or RNA. It also refers to the mating of individuals from two related species.

Inbreeding: mating between closely related organisms.

Incomplete dominance: expression of a heterozygous phenotype that is distinct from, and often intermediate to, the phenotype produced by either homozygote.

Incomplete penetrance: the possession of a mutant phenotype without expressing the corresponding mutant phenotype.

Interphase: the portion of the cell cycle between cell divisions.

Intron: the non-coding portion of a gene, which is transcribed and subsequently removed from the mRNA molecule.

Inversion: the reversal of a segment of DNA.

Karyotype: a display of a cell's chromosomes organized according to size.

Lethal allele: an allele whose expression results in death of an individual.

Linkage: two loci that are situated close together on the same chromosome.

Locus: the chromosomal location of a specific gene.

LOD (logarithim of odds) score: a statistical test to assess the likelihood of two genes being linked.

Marker gene: a gene or short DNA sequence whose chromosomal location is known, which has an identifiable phenotype and whose inheritance can be followed.

Mean: the arithmetic average.

Meiosis: a process of nuclear division during which haploid cells are formed from diploid cells.

Messenger RNA (mRNA): RNA molecules that are transcribed from DNA and translated into an amino acid sequence.

Metaphase: the stage of nuclear division during which condensed chromosomes lie in the central plane between the two poles of a cell.

Microarray: a grid of DNA fragments (typically tens of thousands) attached to a small solid surface which are used as probes for complementary sequences.

Mitosis: a process of nuclear division that produces cells with the same chromosome number and genetic complement as the parental cell.

Monoecious: separate male and female flowers on a single plant.

Monohybrid cross: a genetic cross in which the parents possess different forms of one trait.

Monomorphism: the existence of only one allele at a locus.

Monosomy: an aneuploid condition in which one member of a pair of chromosomes is absent.

Mosaic: the presence in one individual of cells with different numbers of chromosomes.

Multifactorial trait: a trait that is the result of interactions of multiple genetic and environmental factors.

Multiple alleles: three or more alleles of one gene.

Multiplication rule: the probability of the co-occurrence of two or more independent events is calculated by multiplying the probability of each separate event.

Mutant: an individual, or cell, carrying an altered gene.

Mutation: an alteration in gene or chromosome structure.

Narrow sense heritability: the proportion of phenotypic variance that is due to additive genetic variance.

Natural selection: the evolutionary process by which the better adapted individuals survive and reproduce.

Next-generation sequencing: new post-Sanger sequencing methods that achieve high throughput and simultaneously sequence 10^3–10^6 DNA fragments.

Non-disjunction: failure of homologous chromosomes or sister chromatids to segregate into separate cells during meiosis or mitosis.

Nucleosome: a structural unit of eukaryotic chromosomes, formed by 150 bp of DNA wrapped around a histone core.

Nucleotides: the covalently linked building blocks of DNA and RNA.

Oncogene: a growth regulator gene that has mutated, causing uncontrolled proliferation of cells and ultimately a cancerous tumour.

Okazaki fragment: small segment of DNA produced during DNA replication.

Palindrome: a short DNA or RNA sequence that is repeated nearby in reverse complementary orientation.

Parthenogenesis: development of an unfertilised female gamete without fertilization.

Pedigree: a diagram that shows family relationships and transmission of genetic traits over several generations.

Phenotype: the observed characteristics of an individual produced by specific genotypes.

Phosphodiester bond: the covalent link between adjacent nucleotides in DNA and RNA.

Plasmid: circular, double-stranded DNA molecule within the cytoplasm of prokaryotic and a few eukaryotic cells; capable of independent replication and often used as a cloning vector in recombinant DNA techniques.

Pleiotropy: describes genes that have multiple phenotypic effects.

Polygenic: describes a trait determined by the additive effect of many genes.

Polymer: a macromolecule produced by multiple copies of a simple molecule.

Polymerase: see **DNA polymerase** and **RNA polymerase**.

Polymerase chain reaction (PCR): a technique to produce multiple copies of a specific DNA sequence via repeated rounds of primer directed DNA synthesis.

Polymorphism: the existence of two or more different alleles at a locus, the frequency of the most common being 0.99 or less.

Polyploid: an individual, or cell, that possesses three or more sets of chromosomes.

Polysome: a structure consisting of two or more ribosomes engaged in the translation of a similar number of mRNA molecules.

Population: a local group of a single species within which mating is actually or potentially occurring.

Primer: a short DNA or RNA sequence which serves as a starting point for DNA synthesis.

Prokaryote: an organism whose cells lack a true nucleus and membranous organelles.

Prophase: the first stage of nuclear division.

Protein: a molecule composed of one or more polypeptides, each consisting of covalently linked amino acids.

Punnett square: a table that specifies all the possible genotypes that can result from fertilizations between the gametes of a pair of mating individuals.

Pure breeding: see **True breeding**.

Purine: a double ringed nitrogenous base found in DNA and RNA.

Pyrimidine: a single ringed nitrogenous base found in DNA and RNA.

Pyrosequencing: a DNA sequencing method which detects the pyrophosphate as it is produced during DNA synthesis.

Quantitative trait: a characteristic that can be measured on a continuous scale, and whose expression is the result of the additive effect of many genes and environmental factors.

Reading frame: the way in which a nucleotide sequence is read in groups of three nucleotides (codons) during translation.

Recessive: an allele that is only phenotypically expressed in the homozygous state.

Reciprocal cross: a paired cross in which the genotype of the female in the first cross is represented as the genotype of the male in the second cross, and vice versa.

Recombinant DNA: a DNA molecule that consists of DNA from more than one parent molecule (e.g. human DNA inserted into a cloning vector).

Recombination: the process that leads to new combinations of alleles, as the result of crossovers during parental meiosis.

Replication origin: a DNA sequence that signals the start point for DNA replication.

Reporter gene: a gene, with an easily identifiable phenotype, that is attached to the regulatory sequence of another gene of interest to signal when this gene is expressed in a cell or organism.

Repulsion: a term used during gene mapping to describe dominant and recessive alleles linked together on one chromosome.

Restriction enzyme: a bacterial enzyme that cleaves double-stranded DNA at a specific nucleotide sequence (restriction site).

Reverse transcription: the synthesis of complementary DNA (cDNA) from an RNA template.

Ribosomal RNA (rRNA): RNA molecules that, along with protein molecules, are structural components of ribosomes.

Ribosome: the cytoplasmic site of translation of mRNA into amino acid sequences.

RNA polymerase: an enzyme that catalyzes the synthesis of RNA from a DNA template.

Robertsonian translocation: the fusion of the long arms of two non-homologous acrocentric chromosomes at their centomeres.

Selection coefficient: a measure of the relative fitness of different genotypes.

Selfing: reproduction by self-fertilization.

Semi-conservative replication: a model of DNA replication in which each daughter molecule consists of an original parental strand and a newly synthesized strand.

Sense strand: the DNA strand that is complementary to the template strand for transcription. It has the same sequence as mRNA.

Sex chromosome: a chromosome, such as the X and Y mammalian chromosomes, that is involved in determining the sex of an individual.

Sex linkage: refers to genes located on a sex chromosome.

Single nucleotide polymorphism: a single nucleotide variant in a DNA sequence.

Solenoid: a structural unit of eukaryotic chromosomes consisting of approximately six nucleosomes.

Somatic cells: all cells other than germ cells.

Spindle: a network of microtubules that moves separated pairs of chromatids and homologous chromosomes to opposite poles of the cell during mitosis and meiosis.

Sporophyte: the diploid spore producing generation of plants.

Standard deviation: a quantitative measure of the amount of variation shown by a population; it indicates the percentage of measurements within a certain range of the mean.

Sticky end: the single-stranded extension of a double-stranded DNA molecule.

Synapsis: the pairing of homologous chromosomes during prophase 1 of meiosis.

Taq **polymerase:** the heat stable DNA polymerase that is most commonly used in the polymerase chain reaction.

Telophase: the stage of nuclear division during which separated chromosomes or chromatids reach opposite poles and new nuclear envelopes form.

Test cross: a cross between an individual expressing the dominant phenotype, but of unknown genotype, and a homozygous recessive individual.

Tetraploid: an individual, or cell, which possesses four sets of chromosomes.

Three-point cross: a test cross between one individual heterozygous at three loci and a second which is homozygous recessive at the same loci.

Thymine: a pyrimidine base which pairs with adenine in DNA.

Topoisomerase: an enzyme that controls the level of supercoiling in DNA.

Transfer RNA (tRNA): an RNA molecule that brings specific amino acids to the ribosome, as required during translation.

Transformation: the acquisition by a cell of new genes.

Transgenic organism: an organism that has had its genotype altered by the introduction of new gene(s) into its genome by genetic manipulation.

Transcription: the process during which an RNA molecule is synthesized from a DNA template.

Translation: the process during which an amino acid sequence is assembled, at a ribosome, according to the specific nucleotide sequence of an mRNA molecule.

Translocation: the exchange of genetic material between two non-homologous chromosomes.

Triplet: a sequence of three nucleotides which codes for a specific amino acid.

Triploid: an individual, or cell, that possesses three sets of chromosomes.

Trisomy: an aneuploid condition in which an extra copy of one chromosome is present.

True breeding: a variety or strain that yields progeny like itself (homozygous).

Uracil: a pyrimidine base found in RNA which pairs with adenine.

Variance: a quantitative measure of the variation of values from a mean.

Vector: the vehicle (e.g. plasmid, cosmid, BAC, or YAC) used to carry a DNA insert into a host cell.

Y chromosome: a sex determining chromosome; male determining chromosome in mammals.

Yeast artificial chromosome (YAC): a cloning vector which carries large DNA fragments (up to 3Mb) and which includes a yeast centromere, a pair of telomeres and a replication origin.

X chromosome: a sex determining chromosome.

Zygote: the diploid cell resulting from fertilization.

Further Reading

There is a vast and wide-ranging genetics literature to be found, both on the World Wide Web and in print. This list includes a selection of general genetics texts that further develop many of the ideas introduced in this book together with books that focus on specific topics.

Brooker RJ (2014) Genetics: Analysis and Principles, 5th ed. McGraw-Hill.

Brown TA (2011) Introduction to Genetics: A Molecular Approach. Garland Science.

Falconer DS & Mackay TFC (1996) Introduction to Quantitative Genetics, 4th ed. Longman.

Fletcher H & Hickey I (2012) BIOS Instant Notes in Genetics, 4th ed. Taylor and Francis.

Frankham R, Ballou JD & Briscoe DA (2010) An Introduction to Conservation Genetics, 2nd ed. Cambridge University Press.

Griffiths AJF & Wessler SR (2011) Introduction to Genetic Analysis, 10th ed. WH Freeman.

Hartl DL & Jones EW (2011) Genetics: Analysis of Genes and Genomes, 8th ed. Jones & Bartlett Publishers.

Hartl DL (2000) A Primer of Population Genetics, 3rd ed. Sinauer Associates.

Hartwell LH, Goldberg ML, Fischer JA et al. (2014) Genetics: from Genes to Genomes, 5th ed. McGrawHill.

Jorde LB, Carey JC & Bamshad MJ (2009) Medical Genetics, 4th ed. Mosby.

King RC, Mulligan P & Stansfield W (2012) A Dictionary of Genetics, 5th ed. Oxford University Press.

Klug WS, Cummings MR, Spencer CA & Palladino MA (2010) Essentials of Genetics, 7th ed. Pearson.

Krebs JE, Goldstein ES & Kilpatrick ST (2013) Lewin's Genes XI, 7th ed. Jones and Bartlett Publishers.

Lesk AM (2011) Introduction to Genomics 2nd ed. Oxford University Press

Pierce BA (2014) Genetics: a Conceptual Approach, 5th ed. Macmillan

Plomin R, DeFries JC, Knopik VS & Neiderhiser JM (2013) Behavioural Genetics, 6th ed. Worth Publishers.

Primrose SB & Twyman RM (2014) Principles of Genetic Manipulation & Genomics 8th ed. Wiley-Blackwell.

Russell PJ (2011) iGenetics: A Molecular Approach. Benjamin Cummings.

Strachan T & Read A (2010) Human Molecular Genetics, 4th ed. Garland Science.

Strachan T, Goodship J & Chinnery P (2014) Genetics and Genomics in Medicine. Garland Science.

Thomas A (2013) Thrive in Genetics. Oxford University Press.

Index

Printed in the United States
by Baker & Taylor Publisher Services